The True History
of
Religion

How Religion Destroys the Human Race
and
What the Real Illuminati™ Has Attempted
to do Through Religion
to
Save the Human Race

This is the first of three volumes:

*The True History of Religion—How Religion Destroys the
Human Race and What the Real Illuminati™ Has Attempted
to do Through Religion to Save the Human Race*

*A New American Scripture—How and Why the Real
Illuminati™ Created the Book of Mormon*

One People, One World, One Government

The True History of Religion

of

Religion

*How Religion Destroys the Human Race
and
What the Real Illuminati™ Has Attempted
to do Through Religion
to
Save the Human Race*

The Real Illuminati

Worldwide United Publishing
Melba, Idaho

The True History of Religion—How Religion Destroys the Human Race and What the Real Illuminati™ Has Attempted to do Through Religion to Save the Human Race

December 2, 2019
First Edition

SOFTCOVER ISBN 978-1-937390-16-7

Library of Congress Control Number: 2019954654

Worldwide United Publishing
an imprint of Pearl Publishing, LLC
2587 Southside Blvd., Melba, ID 83641
www.pearlpublishing.net—1.888.499.9666

CONTENTS

INTRODUCTION   i

Chapter 1   The End of the World   1
Chapter 2   The Real Illuminati™   12
Chapter 3   The Evil of Religion   26
Chapter 4   The History of Religion   34
Chapter 5   True Messengers   44
Chapter 6   Christianity   55
Chapter 7   Islam   66
Chapter 8   A New American Scripture   80
Chapter 9   Muslims and Mormons   94
Chapter 10   The Mortal Experience   105
Chapter 11   Influencing the Renaissance   115
Chapter 12   The Apocalypse   127

Introduction

Religion is the main reason that the people of this world do not unite together as a human race to help each other and solve humanity's problems. There are few who disagree with this statement on some level. But there are billions who blame the problem, not on *their* particular religion, but on everyone else's. People want to believe that *their* religion is the solution to solve everyone's problems, to solve the world's problems.

Many believe that if everyone else in the world would adhere to what *their* religion said, what *their* god said, then the world would change for the better. There are hundreds and thousands of different belief systems and religions in the world that make the claim. This causes division and many social problems. No belief system has introduced a religious concept or idea that has offered substantial proof that it can save humanity from its own demise.

If we were to explain each of the world's religions, this book could be many volumes. Instead, we have chosen to concentrate our efforts of disclosure on the religions that hold the most power over current humanity—especially the few main religions that establish and promote the economic system that has subjected all of Earth's inhabitants to its power and means. No one can live outside of this system.

Perhaps it would be more relevant to our treatise to describe the economic system that carries humanity on its back as a "scarlet colored beast, full of names of blasphemy, having seven heads and ten horns"—

representing the seven main continents of the world and the ten countries that control all economic activity.

Humanity's problems revolve around economic disparity and inequality. Solving this would greatly enhance humanity's chances of survival. Therefore, the focus of this book will be on the main religions that started the economic structures that control our modern world: Judaism, Christianity, and Islam.

Which will be the first of these religions to step up and proclaim the obvious, everyone is right, which makes everyone wrong? You might believe that *your* religion is the right one, the only true religion on Earth. All other religious believers think the same thing. Who will be the first to admit that *their* religion is actually part of the problem?

Will the Jews be the first to admit that none of the events or characters presented in their Torah are any more factual or real than those created by the ancient Greeks, which are now generally accepted as myths? Will it be the Christians who finally step up? Will it be the Muslims? But would the Christians or Muslims have to step up and admit it, if the Jews did first? Christianity and Islam would not exist without Judaism.

We could have mentioned all of the other major religions in the world. But when honestly considered, the Jews, Christians, and Muslims are those who are currently causing most of the world's problems today.

Why would the leaders of these religions admit something that would destroy their worth to their followers? They wouldn't. If they admitted that their religion was part of the problem, they would lose the donations, praise, and honor that their followers give them.

Most religious leaders in the world "know not that they know not." A few of the more honest ones know that their religion is not reality. But these few are backed up against a potential wall of ridicule and rejection, built and supported by their followers. They realize they would lose everything that supports them and gives them purpose and value, if they admitted that their religion is part of humanity's problems. These few continue to play the role for their own sake. It doesn't matter to them what their religion is doing to the world. Their leadership in religion serves their personal needs. And they can only hope that it serves the needs of their followers.

There's an old saying that we invented and introduced in various ways in different cultures in the past:

> He who knows not, and knows not that he knows not, is a *fool*. Shun him. He who knows not, and knows that he knows not, is *simple*. Teach him. He who knows, and knows not that he knows, is *asleep*. Wake him. He who knows, and knows that he knows is *wise*. Follow him.

Religious and secular leaders, teachers, and scholars (all the "know nots" who think they know) know nothing of substantial worth to solve the problems facing humanity. The more knowledge the people of the world appear to be getting from their religious and secular leaders, teachers, and scholars, the more problems the human race seems to be experiencing. If all of this knowledge was really that good, wouldn't it be making the world better instead of worse for the majority of people on Earth?

This book is not for these fools. This book is for the *simple* and those *asleep*. This book is written in a very simple format. We have made every effort to use simple terms and words to share our message with the world. Sometimes we give the definition of a word or term in parenthesis (like this) so that a common reader might understand. This is to help the *simple*, who might not have the time and means that it takes to learn the meaning of lots of big words.

We write this way to condescend (be disrespectful), not to the simple-minded reader, but to the *fools* who think they are better than others because of the big words that they have learned. These use big words and teach through *superfluous forms of pedagogical gesticulations* (the way "educated" people write to teach their opinion). We do not respect how those "who know not, and know not that they know not" like to use big words, as if their words must be the real facts because of the way they are presented to others. These are the few.

This book was written for the majority of people living on Earth in a way to give everyone an equal chance of understanding its message. There are way too many books written by the few, for the few. Only the few can afford to buy each other's books. Their books are not written for those who cannot afford to buy a book; nor are their books written for those who cannot read. Their books are not written for those who do not understand what their *superfluous forms of pedagogical gesticulations* mean. To the few, those whom they consider to be uneducated and poor readers do not buy their books because they are "uneducated and poor." This is condescending to "he who knows, and knows not that he knows" (those who are *asleep* and need to be awakened).

When the simple-minded read something, and it agrees with their *common sense*, what they read is confirmed and accepted, because it makes sense to them. If they cannot understand the words, they cannot make sense of what the words are trying to say. This *sense*, which should be *common* among us (hence the term "common sense"), is often affected by what each person learns from the time of their childhood.

If the adults of this world returned to the *common sense* that they had when they were little children, what *sense* about things would they have? Can we remember what it was like to be a little child? Can we remember what we knew or didn't know then? None of us knew anything about religion as a little child.

The few might mock the information in this book that we provide to the rest of humanity. This is because it destroys their pretended value in the world for being "educated," as they perceive themselves. They might mock this book because it is free to download and read online, because *they* write books to get money or praise. They might mock this book because no author takes personal credit for the information, or for the success of the book, if any there might be.

There is little doubt that the few fools are going to mock us, the Real Illuminati™, for our claims to be ones who know, and know that we know, and therefore are *wise*. Follow us.

If the *simple* and *asleep* take the time to read this book, they will know who the *fools* are and who the *wise* are. Upon pondering the information given in this book, the wisdom and understanding of the few *fools* who have convinced others that "they know" will begin to perish (disappear). If our wisdom and understanding make more sense than what *fools* teach, especially to the simple-minded, the *fools'* value and worth diminish. The few will not allow their self-worth and value to fade without a fight.

The few pretend to share *their* knowledge with prudence (carefully and cleverly). However, once confronted with Real Truth that makes more sense to the common person, the few will want to hide *their* teachings and *their* understanding of things so that an honest comparison cannot be made. Sometimes information is presented that makes more sense than anything that the *simple* and *asleep* have been taught or believe. If the *simple* and *asleep* confront the person who taught them what makes *less sense* than the new information, the few *fools* will want to hide to avoid confrontation. They will not want to face what they cannot dispute rationally (what they cannot logically argue).

We mock the very reason why most people write books. If a book is not presented as a fictional story, it is written to express the author's opinion as "fact." These are the non-fiction books. They contain specifics that the majority doesn't have the patience or the time to read. The majority are so tired and emotionally drained by the daily grind required of them, that they don't want to think about any heavy information. They want a story.

If presented as a story, a book is usually fiction. Even stories that are "based on actual facts" are often more fiction than fact because they are presented in such a way that it entices a person to read the story and to be entertained.

The most successful books throughout history have always been fictional stories that captivate the reader and do not require a person to think much. These books are successful because they are written as a story, the intricate and difficult details of which must be filled in by the reader's imagination. The more details that are presented in telling the story, the less the reader has to think and the less energy a person has to use to fill in the blanks. These are the most successful books. These are what often turn fiction into "fact."

Research the list of the most successful books in history. All of them are fantasy. The majority of people pick up a book to read in order to be entertained. The "few" pick up a book to learn things that make them *think* they are smarter than those who are just entertained. These books are schoolbooks, scientific publications, and other writings that are presented as non-fiction.

Consider the most successful book in the world. Is it fantasy or fact? We are going to prove to the world—to the few and to the majority—that the most successful book of all time is, in fact (as all other successful books are) a fantasy. But when we mention the name of this book, the reader who believes that the book is not fantasy might condemn us and read no further. Why?

If the book is fact, then *nothing* that can be written or said about it will *ever* prove it to be a fantasy. Wouldn't common sense prevail in deciding? Why would one fear reading any new information about what one assumes is fact?

The fear comes from a person not wanting to be wrong. We value ourselves on what we believe is the truth, especially those truths that define us, that control our lives, and give us personal value and worth. Few will take the chance to be proven wrong.

What if the most successful book in history is the cause of all humanity's problems? Well, not the book itself exactly, but rather the perception (opinion) held by the majority that it is *not* fiction. What if the majority changed their minds and no longer accepted the book as fact, but rather just as fantastical as the books that entertain them? Wouldn't this change in perception help heal the problems that the book caused?

What if the few created the book in order to control the majority? Isn't it true that fantasies and stories can control people? What effect does the idea of Santa Claus "making a list and checking it twice, gonna find out who's naughty or nice," have on a child's mind? Santa Claus is a huge success, not only at controlling a child's behavior, but in marketing and making money for those who promote the idea.

The most successful book in the world is the Bible. It was created by the few to control the rest. It is a fantasy that has been presented by the few as "fact" in order to provoke certain behaviors from the majority. It actually details a list that God checks twice in order to find out who's naughty or nice. If God checks His list and finds out you're nice, you're going to be saved and go to heaven. If you're naughty, you're going to hell (at least according to the stories of the Bible).

Consider all of the books that have ever been written. Consider all of the knowledge that has been expressed (shared) in all of the books that have ever been written. That's a lot of knowledge. But what good has any of this knowledge done for the majority? Sure, it *has* benefited the few, because they are the only ones who read most of it and value themselves based on what they read and memorize. Isn't that all that most college degrees are: the reading and memorization of knowledge that a student gets from a book (or from a professor who got it from a book)?

The world is getting worse for the majority of humanity. But it seems to be going pretty well for the few, at least currently.

Let's give these "few" an actual number. Let's call them the one percent (1%). There are about 8 billion people in the world. One percent (1%) of 8 billion is 80 million people. Honestly, there are probably not 80 million people on Earth who are not slaves to someone else, or rather, dependent on someone else for the basic necessities required for their personal life. How many people are millionaires—financially independent? About 47 million. How many people are billionaires? About 2,600. Let's round off the financially independent to 50 million people. That's a little more than half of one percent (½ of 1%) of the world's population (0.625%)!

We find fault with these few. But the majority should not. How can the majority find fault with the few, when the majority want to be *part* of the few? We fault the few because they have created a society that forces the majority into a unilateral codependency; which means that the majority are subjected to an excessive emotional and physiological (basic needs to live) reliance on the few. The few control the money and provide jobs. Without that income, the majority wouldn't be able to live.

Without the few, it's honestly hard to imagine a world like the one in which we live today. To give proper credit where it is due, the few are responsible for the incredible advancements in technology and innovation. But let's not forget that these advancements benefit the few much more than these do the majority. The few would not have been as motivated if they found no benefit for themselves. Still, how can the majority fault them, when the majority wishes to *be* them? We (the Real Illuminati™) fault them. We have always faulted them, and have done everything within our power to equalize the benefits of humanity's progression and social evolution.

The greatest civilizations on Earth during the past 10,000 years are a result of the efforts of "the few," not the majority. The only exception to this is from the labor that the majority provided, doing what "the few" wanted them to do, like building the great pyramids and skyscrapers, for example.

In the past, civilization advanced and progressed on Earth. Eventually, people discovered the ultimate innovation (best improvements) in politics, business, social welfare, and national warfare (as they supposed). This was known as the Great Roman Empire.

Up to that point in history, the world had never known a united group of people like this "great" empire. Its decadence (excessive indulgence in pleasure or luxury), leisure, and world dominance permeated every aspect of life. Its soldiers meted out death to any individual or to any nation that stood against it. Its sporting events included barbaric acts of competition and brutal human sacrifices. The Romans delivered thousands to their deaths at the fangs and claws of the beasts of the earth. It literally created hell on Earth for the majority in order to entertain the few.

The Great Roman political policies created a very wealthy class of relatively few individuals with great power. This consequently created a middle class of people wanting to *be* "the few." This Middle Class buffered the impact that absolute wealth and power had on the majority of people. This majority included the poor laboring class, who always suffered in poverty.

Throughout history, the poor majority had been able to rise up and overthrow the few in power when the situation called sufficiently for reformation. But when an economic middle class existed, revolution decreased because the complaints of the majority were quieted:

> Teetering on the edge of wealth and believing that they could become just as the wealthy, the middle class stood between the rich and poor: coveting [the rich and] fearing a return to [being poor—thus] validating [supporting] the actions of the rich, while pacifying the cries of the poor.

The military strength of this type of human civilization (the Great Roman Empire, for example) overpowers weaker nations and leaves many desolate. It causes thousands of deaths from the effects of war, hunger, disease, and poverty left over from its military conquests and occupations. There is no part of this type of civilization that promotes equality among the masses, because it wasn't envisioned and established by or for the majority. It existed by and for the benefit of the few.

The United States of America eventually replaced the Great Roman Empire in modern times. The few designed and established it (set it up). Wisely, these few created a Constitution that would always benefit the majority...so the majority was deceived into believing. The few convinced the majority to side with them and fight a war for them.

The original United States Constitution established a Congress of representatives elected by the majority to represent the will of the people. This is a pure form of a *Republic*. At the time the U.S. Constitution was being developed by the people's representatives, it did *not* include the U.S. Senate as part of Congress. This oversight worked well for the majority, but not so much for the powerful few. Once accepted as the law of the land, some framers were afraid that the people's Congress would began to establish laws that would benefit the majority (of a pure democracy) but not sit well with the minority. The laws that would be established by a democratic congress (majority rules) would benefit the majority.

The few had to do something to protect their minority rights and needs. The few could see the threat that a pure democracy would have to themselves. One Founding Father fought to "protect the minority of the opulent [the wealthy] against the majority"!

To compromise, the framers of the Constitution created the United States Senate to protect the rights of the few. From that time forward, no law could be passed by the House of Representatives without the approval of the Senate.

At first, all U.S. Senators were appointed by the rich few. It took over 100 years before the majority uncovered this great deception. In 1913, the American people were finally able to vote for their Senators. The few began to lose their power over the majority. But these few still maintained a great power and influence over the lives of the majority, because the majority still depended solely on the few for a paycheck.

A series of events led to the Great Depression and the greatest social unrest the United States had ever experienced. The U.S. government realized that it needed to do something to maintain its power and control over the people before they began a second revolution to overthrow their current government.

To avoid social upheaval and the increasing social unrest, the U.S. government started providing people with free food, free housing, and free healthcare. But the American majority was a proud majority, especially the people of the Middle Class (even though they were still part of the "poor majority" compared to "the few"). They didn't want handouts; they wanted jobs.

Again, the U.S. Middle Class served the same purpose that the Middle Class had for the Great Roman Empire:

> Teetering on the edge of wealth and believing that they could become just as the wealthy, the middle class stood between the rich and poor: coveting [the rich and] fearing a return to [being poor—thus] validating [supporting] the actions of the rich, while pacifying the cries of the poor.

The few who had control knew exactly what they were doing. They knew how to "pacify the cries of the poor." But now they needed the help and support of the Middle Class, people who wanted to be just like them. They needed the Middle Class to vote for their Senators.

Although their pride keeps the American people from admitting it, they are easily deceived by the powers that control them. Americans need their government. They need the help of the few, because they want to *be* one of the few.

The United States learned no valuable lessons from the fall of its predecessor, the Great Roman Empire. The Romans existed for about 1,000 years before things started spiraling out of control. At this writing, the United States has only existed for not quite 1/3 of that time. The divisions in American society—created by bipartisan (two-party) politics—are the exact same divisions experienced by the Romans before the fall of their great civilization.

There was one book that saved the Roman Empire from complete devastation: the Bible.

The Bible didn't actually save anyone. It was used by the Roman government to control the people who were about to rise up in revolution to destroy it. Using the Bible as its authority, part of the Great Roman Empire was able to transform into the great European nations of medieval times. These European nations also rose to power by using the same Bible as *their* authority.

The Bible also inspired American policy and action more than any other book, more than any other philosophy of man, mingled with scripture. The Bible inspired the creation of a Christian state (a Christian government), even when it was unconstitutional to do so. (The separation between church and state was supposed to be the basis of a righteous constitution.)

In the United States, what branch of government determines if something is unconstitutional or not? The United States Supreme Court. Nine individuals control the lives of the rest of the American people because of their power to interpret and to accept or reject the laws created by the people's representatives.

Consider the Real Truth about how these nine people perform their daily tasks as the most powerful group of people on Earth:

The Supreme Court begins on the first Monday in October and continues until late June or early July. The Justices get salaries, as well as a three-month recess (paid vacation). They only allow verbal argument while the Court is in session from 10:00 a.m. to 12:00 p.m., just two hours! While the Court is not in session, the Justices are supposed to review the cases before the Court and prepare their decisions. Each Justice is allowed to hire up to four clerks, who are the ones who actually do all the work.

Most of the Justices begin their day with a personal prayer in chambers to their particular, personal God. Currently, there are three Jewish and six Christian judges sitting on the most powerful Court on Earth. Their respective religions emphasize the importance of praying daily to *their* God (the God of the Bible), seeking direct inspiration and guidance to help them make the right choice ... according to their interpretation, which is often influenced by their relationship with *their* God.

When the Court is in session, the 10 a.m. entrance of the Justices into the Courtroom is announced by the Marshal. Those present, at the sound of the gavel, arise and remain standing until the robed Justices are seated following the traditional chant:

> The Honorable, the Chief Justice and the Associate Justices of the Supreme Court of the United States. Oyez! Oyez! Oyez! All persons having business before the Honorable, the Supreme Court of the United States, are admonished to draw near and give their attention, for the Court is now sitting. God save the United States and this Honorable Court!

The chant is asking the God of the Bible to save the United States and this Honorable Court. So where is the separation between church and state?

It's this Honorable Court that prepares the list that God checks twice. A person is nice if the person obeys the laws that the Court upholds. A person is naughty if they do not.

These nine people get to interpret which laws are constitutional and which ones are not. These nine people are far from separated from religion. As the chant that introduces and recognizes their authority reveals, these powerful humans represent the epitome (a typical example) of religion: the few making rules in order to control the majority.

But how many people really understand the powers that control their lives? How many understand how laws are made? What they do know is how these laws are *forced* upon them. Individuals either obey the laws, or they are jailed or killed for disobedience. Americans are easily swayed by the powers and laws that control them. They need their laws just as much as they need their religions.

We will reveal in this book how we created a new source of "God's will" (scripture) for the American people in the early nineteenth (19th) century (the early 1800s). We created it to introduce new concepts that might open people's minds and change their views on Christianity. Knowing how powerful the Bible had become, how dependent upon it the majority had become, and knowing that the Bible was fantasy, we created a fictional story for our "word of God," a new American scripture.

It was our hope that the American Christians would relate to the story and consider it to be a story about their present condition. In our story, we tried to inspire the reader to "liken all scriptures unto [themselves], that it might be for [their] profit and learning."

In our new scripture, we made it very clear about what was happening in the modern American legal system:

And now behold, I say unto you, that the foundation of the destruction of this people is beginning to be laid by the unrighteousness of your lawyers and your judges.

In order to avoid the same destruction that the ancient Romans experienced, the people must demand change from the powers that control them, starting with "lawyers and judges." To change any government institution, people must first understand it; and more importantly, understand what motivates and inspires those who hold power in it.

We inspired another ancient quote: "If you know the enemy and know yourself, you need not fear the result of a hundred battles. If you know yourself but not the enemy, for every victory gained you will also suffer a defeat." Through this book, we intend to help the majority, first, know themselves better, and then, know the few, who must be defeated in a final battle in order to save humanity.

Violence has never worked. Revolution has never made lasting changes that benefit everyone. No matter how strong the revolutions, no matter how massive the protests, the majority has failed every time to make this world a better place of fairness and equality for all.

The majority does not possess the physical strength to overthrow the secret combination of political, philosophical (religion and science), and business powers that control them. If the majority can learn anything from history, it is that no revolution, of any kind, of any number, of any strength will be successful at overthrowing the few. This is because when the majority has overthrown the few, they (the majority) reappoint another "few" to start the same cycle of secret combinations all over, again and again. Lasting change will never occur by the sword of physical strength. However, lasting change is possible.

The book of Revelation in the Bible is a disclosure of what we know will save the human race. Explained in this book (*The True History of Religion*), the world will discover that we, the Real Illuminati™, were the ones responsible for the book of Revelation. We will provide the evidence of this throughout this book.

We presented Revelation in religious symbolism, analogous (similar) to Christianity, in order to reach the Christian world. This is because members of this religion control most of the powerful nations on Earth. When the symbolism of Revelation is unfolded, one of the clues we gave will be understood correctly:

> If any man have an ear, let him hear. He that leadeth into captivity shall go into captivity. He that killeth with the sword must be killed with the sword. Here is the patience and the faith of the saints.

The Roman and American governments were not and are not successful at controlling their people through military power. Instead, they were and are able to control their people by another power that is stronger than any physical power on Earth: BELIEF.

We do have something to say that is very condescending about the majority. The few know a lot more than the majority does. They know a lot more about the majority than the majority knows about themselves. The few use this knowledge to control the majority for *their* sake (the sake of the few), believing that it is also for the sake of all of humanity. The only way that the majority will be able to fight the few and make changes that benefit everyone equally, is for the majority to know more of the Real Truth than the few do.

The few do not have the solutions to humanity's problems. Although the few *claim* access to a lot of knowledge (all that has been accumulated during the last 10,000 years of human history), none of it has done humanity any good—none of it.

The world's knowledge does, however, serve the needs of the few at the expense of the majority. The only way to reverse this trend and save humanity is for the majority to have access to knowledge that the few might find hard to accept. The few might find this new knowledge hard to accept because it diminishes their value and worth over the majority.

But knowledge is just knowledge, and there is a lot of it in the world. Intelligence is the ability to use knowledge properly.

We have written this book to give the majority this intelligence.

Chapter 1

The End of the World

A world ended on October 13, 2019. On this day, the planet Fortnite, along with all of its inhabitants, was destroyed.

The millions of humans interacting and participating with each other on planet Fortnite—each doing their own thing, in their own way—saw an incredible explosion. Then everything (including their avatars) was sucked into a big black hole. There had been an environment of land, water, plants, and animals, where these humans interacted with each other on a daily basis. After the explosion, only a black hole remained in a virtual universe that very few of the human players understood.

The humans were emotionally "blown away" at the thought that the world in which they had spent many hours playing during their life, no longer existed. How could these humans be emotionally affected by the destruction of this world, unless they were still alive and conscious of the fact that their planet Fortnite had been destroyed by the creation of a black hole?

Obviously, these humans weren't actually living on planet Fortnite. Each had chosen a personal avatar to live on the planet and participate with other Fortnite avatars. Their real Self actually existed outside of Fortnite's virtual universe in another, parallel one.

While people were having experiences on Fortnite (seeing and hearing things) through their chosen avatars, their actual, physical body was present in another world—in their real Earth world. But while their minds (at least their eyes and ears) were experiencing life on planet Fortnite, there seemed to be no other planet. Their eyes and ears were so focused on what their avatars were experiencing (seeing and hearing), that they became oblivious to what was happening around them in their real world.

All of a sudden, with warning only of an impending "event," the people saw and heard an explosion, and then nothing but darkness surrounding a black hole. Their Fortnite life experience was over. Their avatars were all dead. But their real Selves were still alive. Millions of people, just like themselves, experienced the death of their own avatar. Now, no one was able to connect to planet Fortnite and experience life there any longer.

People were disappointed, at least the ones who were having a good life and winning at the Fortnite experience. The others who were not winning—those who were always getting killed on planet Fortnite—weren't as bothered by the abrupt way that the game ended.

There were millions of other people who wanted to play on Fortnite. Many of these were watching what was going on. These observers realized that if they actually joined in, they would soon be killed by those who knew how to play the game better than they did. To these millions of observers, and the millions of others who were quickly and easily overpowered and killed by the more experienced ones, the game wasn't fair. It wasn't worth playing.

Fortnite is just a game. What is described above is exactly what happened on October 13, 2019. Just two days later, the game's creators released "Chapter 2," and life on planet Fortnite began again. This time though, the playing field was a bit more equal and fairer. This time, so hoped the creators, more people would join in and play the game. The creators released a statement about the new improvements they had made, saying, "More fun, less grind."

Consider the type of computer games that will be played in the future. At this writing, technology is advancing faster than at any other time in human history. Game engineers are working quickly and successfully at finding a way to create a pathway between a game player's brain and the computer's software. Instead of just seeing and hearing what happens in a game, players will soon be able to use their other senses to smell, taste, and feel things found in the game's environment.

Science is exploring the human brain and beginning to understand how the brain can create a very real dream experience. This understanding will eventually lead to the invention of interfaces between computer software and a person's brain that can create an artificial dream inside of a person's mind, a *virtual* reality that is just as real as their own dreams. With this advanced technology, perhaps each game might come with a warning:

CAUTION: Real sensory experience is part of the interactive platform of this game. You will see, hear, smell, taste, and feel things in the game's environment that seem real. No part of this game will cause any real physical damage to your body, or end your life. But all interactions will be felt through your sensory receptors as if they were real. For example, if you get shot by a gun, you will feel the bullet. If you get sliced by a sword, you will feel the cut.

Advanced technology will allow one's "real" reality to be altered for the time that the person is playing the game. As the player's brain interacts within the game, the experience will seem just as real as their life experience is when they are *not* playing the game; or just as real as their dreams seem to be as they are experienced in the subconscious mind. Based on the reality of what is happening with the advancements made in gaming, consider what computer games might be like on Earth hundreds of thousands of years into the future.

Some young Fortnite players were asked how they felt after the game ended in an explosion and they were left without the ability to play the game for a couple of days. Their general response reflected the reason why they started playing the game in the first place: they were bored.

If humans on planet Earth learn how to live together in peace and harmony, where there is no want or need, then humanity could continue into the future, indefinitely. In this perfect human world, each person could experience life according to each of their individual desires of happiness, based on the principles of unconditional free will. However, after living in a perfect world for a while, one could become bored. We would want to play games in order to entertain ourselves and gain new experiences that we otherwise could not have in our unchanging and unchallenging perfect world.

In the world of Fortnite, boredom doesn't exist. Players try to stay alive while everyone else in the world is trying their best to eliminate them. A game that starts out with 100 players will end with only one. The players seek for weapons and resources in order to achieve Victory Royale (in other words, to be the last player standing in each game).

Players may participate in as many as 10–30 or more games in a day, each time with an avatar of their choosing. V-bucks are used to purchase items that create value and individuality, which is an important aspect of the game. The more V-bucks a player has, the more fun it is because there are more options for creativity while playing the game.

The creators of Fortnite observed the inequality of the players' skill levels. They wanted to make the game fairer ("More fun. Less of a grind") to attract other players, especially the millions who were just watching. To change the game, the same people who created the world destroyed it and set up a new one.

Fortnite's creators are actually Fortnite's gods. They created the virtual experience for a specific purpose. They ended it for a specific purpose. We would like to think they ended their Fortnite world for a good purpose for all players—to bring more equality, more fairness, and more fun. But chances are high that they ended one virtual world, from which they were making a lot of money, in order to create another one, from which they would make even more.

The new players who play the game know very little about how the world was created or how to properly play the game long enough without being killed too soon. As a new player starts to play, their inexperience makes them vulnerable to the more experienced players who have the most weapons and resources and, therefore, the most power.

Let's pretend that Fortnite is the only game in town; that it is the best game to play in order to play with the most people and have the best gaming experience. Let's suppose that there is no other gaming platform (no other virtual world) available where a person can join others and spend time interacting to relieve boredom.

Let's say that there are 15 billion people who want to play the game. Most of them are just observers. These remain 'observers only' because they realize that they could never win, so why start to play? They realize that they wouldn't be given a fair chance to live for more than just a few minutes before being killed. This is because of the power and experience from within the game of the few who control it.

Those who created the game are much more powerful than those who control the game from within. However, none of the creators (outside of the game) can do much within the game, if they were to join in and start playing. As players themselves, the creators wouldn't have any more advantage within the game than any of the other players. But outside of the game, as the game's creators, they can destroy the world and create a new one that would attract more players.

The creators know that there are millions of people sitting on the sidelines just observing. Those who are just watching are potential customers. If they chose to play, these people would further fulfill the purpose for which the world was created for the Fortnite creators: to make money.

But what if the game was actually created by the 15 billion people as a place where they could interact with each other so that they weren't bored with their regular life? What if the game was the only game in town, the only way that the 15 billion people could have some fun interacting with each other in a virtual world (the opposite of being bored)?

If the purpose of the game was to engage as many people as possible in the experience, finding a way to make it enjoyable for those just watching would be important. But what if just a few players were spoiling it for the rest? Why wouldn't the 15 billion people want to destroy what they created and start over again?

The Real Truth is ... mortal life upon planet Earth is nothing more or less than a very sophisticated and highly advanced human interactive world. This world is being experienced by each mortal's higher, advanced *True Self.* We are actually a highly advanced human race living in another parallel universe.

There are approximately 15 billion of us living on different planets in a mutually shared solar system. We live in the perfect world for us, which we created for us, according to each of our individual desires of happiness. The mortal experience on planet Earth is the only way that we can interact with each other while living on these different, advanced planets.

Think of it this way:

Let's suppose that all nine planets of our current solar system were inhabited by humans. The planets were created to be new homes for people who once lived together on the same planet called Earth. Advanced science allowed us to create these new planets so that we could spread our population throughout the solar system. Let's suppose that these new planets were highly advanced and provided a perfect life for each inhabitant.

While living on these perfect worlds, we would get bored. We've conquered death and disease. We do not allow plants and animals (nor bacteria or viruses) that might threaten human life to exist on our new planets. Living for a long time among the same people, we learn everything that there is to know about the people sharing our planet. But we don't know much about the people who are living on the other planets.

So that we don't get bored with the same people on our perfect planet, we desire to become involved with those living on the other planets in our solar system. To involve ourselves with these others, and not have to travel to their planet (which is their home, not ours), we connect to an intergalactic Internet platform. (If we visited their home planet, it might disrupt the balance of their society, just as their visiting *our* planet might disrupt the balance of our perfect society.)

This vast Internet platform would be powered by our solar system's sun. We would all have mutual access to this virtual world, to which we could connect equally at the speed of light. Earth would be our Fortnite. The platform in which Earth exists would be the universe that we currently observed from Earth. (However, to our *True Self,* our *real* universe is not Earth's universe. To us, as advanced humans, Earth exists in a virtual or parallel universe.)

We start interacting with each other by playing the game of mortal life on Earth. At first, it all goes pretty well. It starts out equal for all, with plenty of room and resources for everyone. Because the game platform allows free will, the players can create whatever kind of world they want, as long as they are able to convince the other players (or force them) to accept it. Some of the other players learn how to control the game and take it over. Soon, the world becomes corrupted by these few, who eventually take over the world.

Now the Earth experience that started out equal for all becomes one where people are dependent on others for their survival. No advanced human can connect to the corrupted game and remain alive without another, more experienced player helping them learn how to play. A player can't just connect to the Earth experience with a game avatar of their own creation, one that looks like them as an advanced adult. The game has changed drastically from what it was in the beginning.

Now, in order to connect to the game, an advanced human has to wait until a mortal body has developed inside the womb of one of the other players. They have to be "born" into the world before they can start playing. Once the new body is fully formed and functional, the first breath an infant takes becomes the connective protocol of (way to participate in) the mortal experience upon Earth. (Some call this the "breath of life.")

This mortal experience is nothing like it was meant to be, as Earth's creators intended for it to be, in the beginning. Earth's creators do not want it to end. But if the players inside the game do not do something to improve the playing field and return it to how it used to be in the beginning, the world will not fulfill the measure of its creation. If Earth does not provide the experience it was meant to provide for those who want to play the game, then it will be destroyed by the "gods" (advanced humans) who created it.

WE (each and every one of us—YOU included) created Earth. WE are these gods. Earth will be destroyed by US—the advanced human beings who want to play the game of mortal life.

Life upon Earth has been going on for billions of years. During those years, our *True Selves* (who we really are) have tried everything that can possibly be done to convince our mortal avatars to start playing the game as it was meant to be played. This was all done according to the laws of nature (the rules of the game). We have failed.

Now the rules must be bent as a last resort.

Our mortal avatars were not supposed to know and understand that they are not the real person behind their existence. Because if mortals actually *knew* this, then at any time they experienced an uncomfortable situation that made them unhappy, they would end their own existence.

You will *know* this Real Truth when you die. When your mortal life ends, no matter how it ends, you will immediately be aware that you continue to exist. You will recognize that your *True Self* is an advanced human of incredible intelligence and individual power (free will). You will then know that YOU are the only god that has ever existed in your experience.

You will realize that all the times upon Earth when you felt the presence of "God" or your "conscience," you were actually feeling the connection that your mortal avatar had with your *True Self.* You will know these Real Truths about your own Self, but also about every other human with whom you shared your earthly experience. You will know that each person who lives on Earth is actually connected to *their True Self.* You will know that there is no other god outside of YOU, outside of EACH of US.

At that time, you will recognize exactly <u>what</u> has corrupted the mortal experience and destroyed humanity. You will understand that, unless confronted and eliminated, this culprit (this CAUSE of the problem), will destroy our Fortnite—end our lives upon Earth.

This culprit is religion. Religion lies to us and tells us that we are not our own god, but are subjected to (or enslaved by) the god that religion has invented to control our actions. Religion teaches us to not depend on ourselves for our own salvation. Religion teaches us that we must depend upon "God," even if this god allows us to experience such a miserable mortal life.

We are taught about this God by religious leaders who personally benefit from the lie. If we were convinced that there is no god other than US, outside of the "god within," then we could unite and do what we must do in order to save Earth and all of humanity. Thus convinced, religion would no longer exist, and those who promoted religion for their own gain might cry out, "Oh mountains and rocks, fall on us, and hide us from this Real Truth!"

Chapter 2

The Real Illuminati™

Our *True Selves* do not want this game to end. Everything that can possibly be done at this time to save it is being done. One of the rules of the game is the support and protection of free will. No one outside the game has the right to enter the game differently than any other player. If it *were* allowed, the game wouldn't be fair to the other players.

In order to change the game from within, there need to be players, within the game, who understand what it is, who understand the Real Truth about human existence. In order to allow a few players to understand things that are against the rules for everyone else to understand, the rules had to be bent, but not broken.

As a person connects to the mortal experience through an infant, no one's brain is capable of remembering anything past what they began to experience after taking their first breath of life. Throughout their life, they cannot remember any life experience other than the one they are living. Therefore, a person's knowledge during the game is limited only to their mortal experiences. This is how the mortal brain is supposed to function.

If a person born into this world could remember who they really are, as one exists as an advanced human *playing the game of mortal life*, suicide would be the solution to end any temporary suffering. Few would last long in the mortal experience, and the purpose for life on Earth would lose its significance.

The mortal brain is not able to bypass the connective power of one's *True Self.* This is because the energy level (the power) of your advanced brain repels the reduced energy level of your mortal brain and keeps it at bay. The only exception to this is if the advanced brain is somehow physically altered to allow a connection. Think of it this way:

Think of the energy that the brain produces <u>like</u> you would think of air pressure. An advanced brain has a higher level of pressure (energy). A mortal brain has a lower level of pressure (energy). If your advanced brain connects to your mortal brain, it is impossible for the air (or the energy) from a lower pressure chamber (your mortal brain) to enter where the higher pressure exists (into your advanced brain).

In order for the lower air pressure to enter a chamber that contains a higher level of air pressure, the air pressure in the higher chamber (advanced brain) must be lowered, or the air pressure in the lower chamber increased beyond that of the higher-pressure limit. (A mortal brain does not function with the same capacity as an advanced human brain. Therefore, it is impossible for the mortal brain to increase its pressure, hypothetically, in order to sense the energy of one's advanced brain, or in other words, know what one's True Self knows.) This can only happen when the rules are bent for a few, but not for the normal mortal person.

The mortal brain (chamber) is only capable of a certain level of energy (like air pressure). The advanced brain (chamber) that is connected to the mortal chamber must be lowered in order for the air to pass into the higher-pressured chamber. Again, this only applies to those of us within the game who understand the Real Truth about human existence. Currently, there are very few mortals on Earth with this cognitive cerebral capacity.

The laws of mortal human nature on Earth do not allow the creation of a chamber of a mortal brain that can handle the higher pressure. Earth life was set up so that the higher pressure from the advanced brain could leach into the mortal brain of a lower pressure, but not vice versa. For example, the air pressure inside an inflated balloon naturally rushes *out* of the balloon into the surrounding environment, but not the other way around.

In order for a mortal brain to know and understand things that an advanced brain knows, one's *advanced* brain must be physically altered, because the mortal brain always remains the same according to the laws of nature. The higher pressure of the advanced brain must be lowered in order to allow the air pressure (or energy) in both chambers to equalize. Once equalized, the mortal brain has complete access to what the advanced brain knows. This is the reality for us (the Real Illuminati™).

Knowledge is nothing more than energy gained from life experience from when we act in an environment and receive energy from that environment—which we then store in our brain as memories, as experience. We see, hear, smell, taste, and touch things in the environment in which we live. When we do this, we transfer energy into our brain.

When we see something, actual energy from the light of the sun enters our eyes. The energy from the sun creates the image in our brain. Or, in other words, the sun transfers some of its *own* energy into our brain to create the energy that causes our brain to "see." We then store this energy in our brain as a memory. In this way, <u>all knowledge and experience is energy</u>.

We—those who are now being revealed to you as the Real Illuminati™—have normal mortal brains. It is our advanced Self's brain that has been altered. We have access to the knowledge that each of our individual *True Selves* have, which the normal mortal person does not.

We entered the game just like any other mortal (at least most of us did). However, because our mortal avatars have failed to fix the problems on Earth, according to the set rules, something had to be done. A rule had to be bent in order to allow a few to know the Real Truth of all things. This is so that we could teach the Real Truth and what needs to be done to those who *don't* have access to this information.

The rules of the game (hypothetically) require the support and protection of free will. Therefore, we cannot impede or violate the free will of others. We can only do what other mortals allow us to do. Even if we try to tell other mortals what we know, if what we tell them feels like their free will is being violated, when what they have chosen to believe is being threatened, they will reject what we tell them. This has been the case throughout most of Earth's history.

We are an esoteric group of individuals (known about or understood by very few people). We work privately behind the scenes of society. We counter superstition (an irrational belief in things such as magic or luck, arising from ignorance or fear). We vigorously challenge religious influence over public life. We strongly oppose the abuses of government power. We disagree with obscurantism (the practice of deliberately making things more confusing or complicated so that people do not discover the truth).

We exist with the intent to influence critical thinking and free thought in the vigorous effort against corruption, when the acts of authority (governmental, religious, traditional, or cultural) inhibit individual free will.

We are not known or meant to be known by the general populace. This is to protect our existence from the societal or cultural authorities who control human life upon Earth, who control you.

If we were publicly known, each or any one of us would be propagandistically branded a threat to society. This means there would be all kinds of false information spread about us throughout the media, throughout your leaders, throughout those to whom you look for your understanding and for the "facts." In so doing, we would be removed, either by incarceration or assassination (imprisonment or death), but at the very least, by alienation.

We have introduced ourselves as the Real Illuminati™. Our name has been perceived to represent a secret society of people working behind the scenes to accomplish something nefarious (evil), something immoral that does *not* benefit humanity. However, the opposite is the Real Truth about us.

We are trying to help humans live together upon Earth in a cooperative society of its most perfect form. Human cooperatives (societies) begin as individuals join them, or are forced to join them, sometimes because there is no other option. Humans seek out the group that best serves their individual needs, whatever these needs may be.

Humans are social creatures. Humans seek happiness. We seek happiness by our associations with others, choosing which society serves our individual needs.

Throughout the history of the world, we have done everything within our power to promote a proper existence for all humans. A perfect individual human existence (life upon Earth) is when a person can exercise unconditional free will that serves the physical and emotional needs of that person, without impeding or affecting the free will of another.

Nothing that we (the Real Illuminati™) have done, are doing, or will ever do, impedes the free will of another person. We do not force, we do not obstruct, and we do not do anything that violates <u>any</u> person's free will. If we did, we would be acting in direct opposition to *why* we do what we do, which is protecting and supporting free will.

If the rich are in power, we do all that we can to work towards a more perfect life without threatening their riches or power. If the poor rise up and threaten to unite and take the power away from the rich, we do all that we can to help the poor devise a different way that will not affect the free will of the rich.

With our help, we have been able to assist the people of Earth to create societies and cultures in the past, which have moved forward progressively to serve and protect individual free will. We have done this by providing the knowledge and understanding necessary to correctly influence governments and ideologies. Our efforts have never lasted for a long period of time, and we are eventually cast out and rejected. We have witnessed many cultures, civilizations, and societies upon Earth end in disaster, usually in complete annihilation, except for a few outcast enclaves that survived.

Our greatest enemies have always been and still are: religion, patriotism, nationalism, and human pride and ego.

The protection and support of individual free will is the overall goal of our group. Nothing is more important to human existence than individual free will. When the actions of an individual or group within a society or culture begin to threaten human life and/or individual free will, we do everything within our power to counter such threats.

We are not magical. We are not superhuman. Our weapons are knowledge of how things really are and of how things have really been in the past, and the intelligence to know how to use this knowledge properly to serve humanity. This knowledge provides us with a clearer picture of how things are really going to be in the future. We call this knowledge: Real Truth. We share our intelligence in hopes of setting humanity on a proper course that will guarantee its survival.

There is a big difference between knowledge and intelligence. You can know a lot of things, but not really understand how to use this knowledge properly. Our motivation, purpose, and strength come from our understanding of how to use the things that we know. Everything that we know centers around individual human existence and the ability of each person to exercise free will in the pursuit of happiness.

Throughout history, the secret combination of political, philosophical, and business powers has greatly diminished the ability of a person to pursue happiness according to that person's own desires. Wherever and whenever this secret combination has existed, so have we, in some form or another, but we are forced to keep our identities secret.

Unlike these secret combinations, we eagerly spread the Real Truth for free without any single member of our *real* group taking personal credit for the incredible truths that we reveal.

The Real Truth can destroy the power that these secret combinations (these "purveyors of injustice") hold over the general public, hold over YOU! These secret combinations withhold the Real Truth and deliberately prevent the facts or full details about what they do behind the scenes from becoming known. (This is obscurantism.)

The masses need to know and understand the way that these secret combinations gain and hold their power and control over the majority. They need to know what is happening behind the scenes. If they gain this knowledge, they will be able to take back their freedom and free will. This will allow the masses to be in power and control—a power and control of the people, by the people, and for the people.

As this combination of philosophical, religious, and economic power proceeds in the dark, we fight to bring their secret acts to light. Currently, these purveyors of injustice hold power over the minds and hearts of the masses. They use this mind control to get the people to do their bidding (follow their orders or keep their "commandments").

You might think that no one controls your mind. This might be a sense of individual pride and self-worth to you, that nobody controls what you think. But this is a deception. It's such a strong deception that you'll never learn the Real Truth until you can get over it. If you are honest with yourself, think about what you believe. Think about what you "know." Your mind is being controlled by those who put the information (knowledge) into your head.

Unless what you think and what you believe are completely original—unless they've never been thought of before—someone in your past has used control over your mind to introduce those things to you. Do not deceive yourself into thinking that nobody controls your mind. Think carefully about this.

Because it is our work to reveal many of the things that are done in secret, those who act in secret use their power over your mind and your heart, to cause you to fear our name. We are the Enlightened Ones, the Illuminati, albeit the Real Illuminati™.

We are the *Real* Illuminati™ because we do not try to hide the Real Truth. If possible, we would shout it from the housetops!

Other false claimants (*false* Illuminati) pretend to be enlightened and value their secrecy. These poison our name. The *false* Illuminati have secret societies and secret initiations, and are no better than the purveyors of injustice that we seek to uncover.

In fact, many of these secret false ones are powerful members of society—they are politicians, philosophical leaders of science and religion, and business leaders. These claim that their works done in secret hold society together; that without them, society would fail. Yet, society has failed, and is failing. The *false Illuminati* are just as responsible for this failure as those we seek to expose.

When we speak the Real Truth openly, these false ones rally their corruptible troops. From the media that they own, and from the pulpit where they spew their fabrications and control your mind, these false ones bolster and excite their "troops" to fight us.

We, the Real Illuminati™, also rally our own troops. But our numbers are few ... very few. We have a mantra that has been passed around in many different ways. This rallying cry inspires us against our enemies, and outlines the protocol by which our True Messengers operate. We had our hand in contributing many things to the story from which this mantra is taken:

> Fear them not; for there is nothing covered that shall not be revealed; and hid, that shall not be known. What I tell you in darkness, that speak ye in light: and what ye hear in the ear, that preach ye upon the housetops.

It is important to keep in mind that everything that we have done, everything that we do, and everything that we will ever do protects the free will of the individual in a peaceful way. Although many of our messengers have been killed for sharing our message with others, we have never used violence towards another. We have never supported or influenced the use of violence towards another.

Instead of the need to defend ourselves, we do not place ourselves in situations in which we might need to. We assimilate our presence into the society in which we operate. We choose a society based on the desires of the people. We seek out a society where the bulk of its population is seeking for the same goals as we are.

In an effort to create or preserve peace throughout history, we have helped certain societies defend themselves against aggression. This was done by giving certain craftsmen, scientists, or researchers ideas that spark technological inventions. These inventions have led to weapons that the aggressor does not possess.

Our intent in providing these types of advanced weaponry and knowledge was to dissuade the aggressor from continuing an offensive attack against another society. We (the Real Illuminati™) do not support vengeance (retaliation) or a vindictive offense (purposefully hurting others), regardless of the amount of destruction and misery the aggressor might have caused.

We were responsible for providing the scientific knowledge to the right people in order to place the control of nuclear energy (atomic power) in the hands of the United States of America. This was to stop the rise of Fascism (extreme oppression) in Europe. When the German nation exalted the white race above all other races and began a systematic massacre to eliminate individuals, it needed to be stopped.

European scientists held most of the keys to modern nuclear technology. These keys were available to the Axis powers (the nations that fought in World War II against the Allied forces), but there were certain scientific principles that they lacked. We provided the rest of the needed nuclear intelligence to America and its allies. This was done to stop Germany and other malicious (cruel) European countries from succeeding.

Our intent for sharing this knowledge and intelligence with the United States included the hope that the United States would retain this incredible new power for the purpose of protecting and supporting individual human rights.

Our attempts to counsel and influence the United States government to protect individual free will have generally failed. The United States shared its nuclear technology information with other countries that were not established on the concept of unconditional freedom in the pursuit of individual happiness. This mishap led to further instability throughout the world, a mishap that we are obligated to try to correct.

We have a warning for the people of Earth:

You must unite as one human race and begin to treat each person equally (alike) and with fairness (equity). You must end worldwide poverty and provide the basic necessities of life to each human, free of charge. IF you do not do this, we (our *True Selves*) are going to allow this solar system to be destroyed.

This world is no longer moving forward. It is progressively moving backwards. Our last and only hope is to provide the people of Earth with the knowledge and the intelligence necessary to stop the world from going backwards, to help propel it forward.

We intend to share our knowledge, and more importantly, our understanding, of the past and present, so that the future can be salvaged.

If the world rejects this knowledge and continues on a slippery slope of human degradation, individual slavery (the inability to exercise free will), ethnic, religious, and national division, our mutual *True Selves* are going to allow this solar system to be destroyed.

In a few basements throughout this world, very young, ingenious, yet inexperienced, scientists are experimenting with nuclear fusion—the power that created our sun. If our *True Selves*—to which a mortal transfigured brain can have full access—find no more use for life upon Earth, then one of these young scientists' *True Self's* brain will be transfigured just enough to allow the boy to access the missing keys needed to replicate the creation of a sun. Upon his experimentation, a new sun will be created that destroys this solar system.

A logical question might be asked about why we claim to have the authority and power to end human life in this part of the universe. We do not need to claim that we have the power to create a new sun, because it is not a claim. Our *True Selves* already possess this knowledge and will only use it similarly to how, and for the purposes for which, we used the same knowledge to help the Allies win the last world war.

The same source (our *True Selves*) that provided us with the knowledge and authority to share nuclear technology with the Americans, will provide a young scientist with the necessary understanding of fusion needed to create a new sun.

We are beholden to the source of this knowledge (obliged and bound to it). It is not a supernatural, magical, or divine source. It is a human source. We hold this knowledge carefully and responsibly. Our *True Selves* will only use it to end this solar system for the same reason that it was created in the first place: to serve the needs of the individuals who live here, who are the ones who also created it.

One of the needs mortals have during this experience is to understand who they are and why they exist. Religions are formed as humans attempt to answer these questions. We have tried to influence societies by providing True Messengers who can provide the proper religion, one that supports free will.

But we have learned from sad experience how easily our good intentions have been corrupted and ignored by the people who receive our message. The original message given to the people by our True Messenger, if not changed by the people, would fulfill the purpose for which we shared it: to bring peace, prosperity, and happiness to the human race.

Religion has caused most, if not all, of the inequality and misery upon Earth. Religion has corrupted the earth experience and changed it from what it is supposed to be. Religion has the most powerful effect on the human mind, which leads to how humans act. Because of all of this, we had to use religion against itself.

Everything that we have done through religion was done with the hope that we could help mortals change the playing field and make life on Earth fair and equal for all people. Because, as we have explained, if we fail to do it ourselves, if our mortal avatars cannot accomplish this, those who created Earth will destroy it and start again.

We have the knowledge to help humanity. We will tell you, not only what needs to be done, but how it can be done. We will give the people currently living on planet Earth all the details that they need in order to improve life upon this planet, as it was meant to be by the creators of the *Game of Mortal Life*.

Until our *True Selves* have given up on our mortal avatars, we will continue to do everything we can to help save humanity. To do so, we must confront religion.

Chapter 3

The Evil of Religion

Nothing has affected the ability of humans to live together upon Earth in peace and harmony more than religion. More humans have been killed, maimed, and persecuted for the cause of religion than for any other reason. Although political and economic factors have added to human misery, religion has done the most damage.

Many religions claim that *theirs* is the best, that *theirs* is the right one; and that if all the people in the world were to adhere to and follow *theirs*, heaven on Earth could be established.

Countless books and studies have been written and published about religion, by the religious, and its effect on human society throughout the course of Earth's history. These publications however, leave out many of the negative effects of religion.

<u>Nothing</u> has affected human thought and action more than religion. <u>Nothing</u> has caused more heartache and pain than religion. Religion has caused innumerable and incomprehensible wars, which have resulted in the loss of an immeasurable amount of human life. Religion causes intimate personal depression that results in emotional guilt, misery, and uncertainty about one's self-worth and value.

In spite of all of the empirical and clear evidence of religion's negative effect on humanity, nothing is more powerful and nothing is more widely employed in the pursuit of happiness than religion. The human search for happiness and value is the core fuel that powers the course of human events upon this earth. And it's in this search for happiness and value that religion performs its greatest role. Religion has centralized itself in the energy of the human mind. It has become the most effective and powerful use of this cognitive (thinking) energy today.

Religion is not only defined and limited to organized belief systems that are generally classified as the world's main religions. Besides the religions of Christianity, Islam, Hinduism, Buddhism, and other prominent "isms" and their offshoots, we must include the new age of spiritualism, psychics, channelers, astrologists, and others engaged in activities focused on the search for happiness and self-worth.

Not to be outdone by these popular and widely spread traditional movements, science is also a religion. Like religion, mortals often use science in their search for happiness and value. Science focuses on a perception of understanding versus the negative view of being ignorant, as scientists often view those of religious belief.

Scientists and researchers become "published" gurus in their own right. Famous scientists throughout history have become science's prophets, its revelators, its seers and ministers. Science quotes its own and accepts its own theories and postulations (assumptions). Science does this with the same affection and honor by which other religions accept their prophets and teachers and quote from their own scriptures. Scientific publications are its scriptures. Scientific theory and religious belief are semantically intertwined (the fundamental meaning of both are woven together as one). Neither one can provide answers with absolute certainty to human's questions. Neither can deny the results of their failure to provide Real Truth and common sense solutions.

Both science and religion provide their followers with certain emotional values of the illusion of happiness and self-worth. For instance, whether receiving a college degree or receiving a certificate of baptism (thus being "saved"), both provide an emotionally-charged feeling of great accomplishment. In fact, these "rewards" are valuable to self-worth and happiness. But in Real Truth, neither is worth any more than the paper upon which the degree or baptism is proven.

Both religion and science can be properly defined under the general term of Philosophy. "Philosophy" is best defined as the study of the fundamental (basic) nature of knowledge, reality, and existence (the very core of them). We can refer to both religion and science, and all philosophy of every nature and kind, as "the philosophies of men."

The male gender has been chiefly responsible for the study of human nature, and men the main originators of both religion and science. Women should not take offense to this statement, because *nothing* the philosophies of men have created on Earth has led to lasting happiness and individual self-worth. On the contrary, when the Real Truth is revealed about how the philosophies of men first started and why they were supported and perpetuated by men, women will gladly disavow their role in the historical development of religion, science, and philosophy.

The philosophies of men mingled with "their scripture" are responsible for the devastating amount of human misery and tragedy to which the human race has been exposed throughout the history of this earth. Who can say which has caused more damage to the pursuit of happiness and self-worth: The belief in a jealous and vindictive God, who only saves His own people and destroys everyone else who does not accept or believe in Him? Or the atomic age with its destruction, fear, and paranoia?

From the relatively little-known scientific "scripture" written by Robert Hooke and Max Planck to the more popular "prophecies" of Einstein and Oppenheimer, the first atomic bomb was created. This has destroyed many lives and has threatened the majority of the human race.

Not to be outdone by nuclear threat is the damage religion has created by the writings and actions *attributed* to the prophet Muhammad of the Islam faith and to Pope Urban II of Christianity. The Muslim conquests and Christian Crusades started "holy wars" (jihads) that have created both internal and external hatred between humans that linger even to this current day.

Human nature is responsible for these philosophies and the tragedies that they create. From the first time that mortals began to wonder about their existence and call upon an outside, invisible, unknown force to answer the questions about their reality, their innate human nature has deceived and misled them. The lingering questions of who we are, why we exist, where we came from, and where we go after we die, circumvent all other emotional activities of the human mind.

Humans have always struggled for the necessities of life, competing with each other for the limited resources of the earth. But this does not compare with the frustration they feel in their mind to find answers. The struggle *within* effects our emotional happiness the most. If one doesn't know the answers and cannot find them from within, it suggests that a person is flawed intellectually or implies that they are uneducated or ignorant in some other way.

Nevertheless, the questions still linger and the search for answers remain foremost in the mind. Is there someone else who might know the answers? Does there exist a messenger sent from the source of all knowledge to provide the answers? "If I don't know, then who does?"

Humans fight the thought that they might be stupid. They become discouraged when they can't find the answers themselves. A battle begins to take place in the mind. This enmity, this hostility, happens because each individual is connected to their True Self. There is a constant battle between the conscious mind (mortal Self) and the subconscious mind (*True Self*). Each person *feels* that they are just as good as everyone else. If they feel they don't understand something that someone else claims to understand, the human nature of equality compels their search for a better source of knowledge and understanding, or causes them to accept another's claims. This is how religions are formed.

Humans know that they are different from any other animal upon Earth. Subconsciously they recognize this difference, but they don't understand *why* or *how* they became so different. This inner battle, this opposition in one's mind (a "cognitive dissonance") compels them to figure it out.

In the search for the answers, the sincere and honest people (those who admit their lack of knowledge) become susceptible to the philosophies of men mingled with scripture. (This means that they become vulnerable or open to other people's opinions and ideas, as taught in religion, science, and philosophy.) This is because humans place an important value on knowing and having the innate questions about their existence answered. Because of this, when one can't figure out the answers for oneself, unscrupulous men (mostly men, some women) profess to have the answers for which the one is seeking.

Knowing of the intrinsic value and continual longing for answers, these false messengers set the price for the information they pretend to have. The one who provides the answer is allowed to bargain with the value of knowledge. And like everything else in a world in which "you can buy anything for money," the searcher will pay almost any price for the answers. They only need to be convinced that the answers are of worth.

The answers to life's questions have become a commodity that one can purchase, sell, trade, and negotiate for in other terms. The unscrupulous ones (those who are dishonest and corrupt) have become the merchants (the sellers) of knowledge. Because of the natural enmity (the battle for self-worth) within the human mind, these merchants do anything they can to protect the pretense (faking it) of their possession of knowledge. One way that they protect themselves and the group they have deceived into accepting their claim of knowledge—a group that gives them a lot of value—is by taking the treasures of the earth (money) they receive from the deceived group and financing armies to reign with blood and horror thereon; which is another way of explaining how people protect their family, community, and nation.

None of the gods created by the philosophies of men mingled with scripture are real. But their power over the human mind certainly is! From the moment that a sincere inquirer of truth, in submissiveness, kneels upon the earth and asks for guidance, "Oh God, hear the words of my mouth; Oh God, hear the words of my mouth. Oh God, hear the words of my mouth," that one becomes vulnerable to being answered by the only "god of this world": one's own mind.

But when the answers do not come from within oneself, the sincere seeker looks elsewhere. No matter how sincere and humble the person is (recognizing their inability to find the answers), they are forced to look for someone outside of their own consciousness to bring them the answers: "I am looking for messengers."

Once open to the possibility that another human just might have the answers, the seeker of truth is presented with various ministers of knowledge, who provide answers made up of the "philosophies of men, mingled with scripture."

Unfortunately, most people fall under the spell of these false messengers, who don't have any better understanding of the Real Truth than the seeker who pays them for an understanding. The false messengers' pay comes in a wide variety of forms. Most receive actual money or other relevant material goods for their services. But ALL receive a payment that is more enticing than all the money in the world: power over another person, respect and standing, and accolade-laced value from their followers.

No religion has ever been able to establish peace and harmony on Earth. Religion has an excuse as to why it has never accomplished this. Each places the blame on those *outside* of *their particular* religion: Gentiles, infidels, sinners, and those whom they consider lost or who refuse to join them. Their excuse is that, if everyone would just join their religion, the world would be saved.

Three of the world's main religions: Judaism, Christianity, and Islam, claim that no matter how hard each tries, none will ever be successful at creating peace and harmony on Earth until *their* particular Messiah comes to Earth (or returns to Earth) with the power and authority to create *heaven* throughout the world.

None of these religions believe that mortals can do it on their own. Ironically, these three religions, more than any of the others, have been the main cause of most of the carnage and death throughout modern history.

Consider for a moment what would happen to these three religions if humankind actually solved all of its own problems without the help of a Messiah. Or perhaps, maybe, there wouldn't be any problems if these three religions didn't exist. If we were to solve our own problems, not only would these three religions not be needed, but their identity and worth would be greatly diminished, as well as the self-esteem, self-worth, and self-value that their members gain from belonging to them.

Returning to science, hasn't it also produced its own power, its own god? "What about cars, computers, cell phones, modern medicine, and the Internet?" After all that science has accomplished, after all the advancements in technology and innovation, has the world become a better place to live? Maybe for the few. Maybe for the ministers of knowledge who have deceived the rest of the world. But not for the majority of the human race.

So, we must ask ourselves a very important question: if religion and science are not the solution, could it be that these 'PHILOSOPHIES OF MEN MINGLED WITH *THEIR* SCRIPTURE' are actually the CAUSE of all human misery? The answer is, YES!

Chapter 4

The History of Religion

When *did* religion first start, how did it start, and why did it start?

In our quest to change the world, we (the Real Illuminati™) have influenced many people throughout history to think beyond the limits of the worldly knowledge that was made available to them after their birth on planet Earth.

Humans are born *without* worldly knowledge. Your knowledge is not of your own making. You gain your knowledge from your parents or from whomever your parents allow you (as a little child) to learn knowledge (such as teachers and church leaders, for example). You learn knowledge from the teachers to whom your parents send you to learn.

Your personal knowledge is restricted and limited by the honor and respect that you give to your parents. As a little child, you are offered rewards and threatened with punishments depending on the respect that you give to your parents' "commandments." If you learn your parents' knowledge and do not question it, you are rewarded with their love and praise. If you question your parents, you are punished with their rebuke, disappointment, and sometimes anger.

You are punished, not for wanting to learn more, or for learning different things than what your parents know, but because your parents feel devalued and disrespected by you. In almost every case, your parents created you for *themselves,* not for you. They have their own reason why they created a child in the first place. A parent's expectations for their child have nothing to do with what the *child* expects out of life. Having children is all about what parents will receive *from* their children.

Parents teach you all that they know. But if your parents' knowledge is not enough for you, and you seek for more or for different knowledge than what they have given you, your parents' value and worth, in their eyes, is threatened. No parent wants a child to see them as stupid, or less informed than their children are.

Before there were books and schools, before there was writing and arithmetic, whatever the parent knew was the complete source of a child's learning. How could members of a family or community learn something if no one else knew it? How much did our ancient ancestors actually know?

Anciently, when a child was born into a family, the child was a valuable asset to the community. As explained, ancient children also served the needs of the parent. New children were needed to help feed, support, and protect the parent-based family unit. A woman desired a man whom she believed was most capable of providing food, support, and protection for her and her children.

As human families encountered other families, they learned that cooperation with each other led to more food, more support, and more protection than could be obtained by a single-family unit. It was much easier to protect human life from a pride of lions, for example, if there were 20 men with spears rather than just one. Humans learned that cooperating with each other in communities was to their advantage.

Humans developed these communities based on their physical similarities. White-skinned people recognized physical resemblance in other white-skinned people. This made them feel more comfortable and more connected. To white-skinned people, darker-skinned people didn't seem quite right. Soon, the whiter races united together to protect themselves against the darker races, and vice versa.

As these ethnic-based communities developed throughout the world, a united *human race* did not exist, nor did it seem possible. At that time, the only knowledge available to the child—received from the family and community—was that the color of human skin was important to self-preservation. Many years of these ethnic divisions created the misguided belief that one color of skin was preferable over another.

The white male living in the north went many months without being able to do a lot of physical work because of the cold weather. He did not develop as physically strong as his darker southern counterpart did. Darker men worked all year long and developed some physical attributes from physical labor (hunting, gathering, protecting) that white men of the north did not develop during the colder winter months.

To counter this physical inequality, the white families of the north united in greater numbers than the darker families of the south did. White-skinned cultures grew more rapidly and became more cooperative with each other than the darker-skinned races.

Once isolated in white-only communities, there were still men who were physically less attractive to the women, and generally less important to the community. These physically inferior men could not hunt and protect the community like the stronger males could. While the stronger men hunted and protected the community, the weaker men were left behind to do whatever they could, albeit in a less important supportive role. They wanted to prove their importance to the community and to the women.

So the weaker men learned to organize supplies for the other men who would go hunt or fight other tribes (communities). When the hunters returned with food, the weaker men accounted for the food and made sure that it was equally distributed to the rest of the community. When the stronger men left to fight, the weaker men made sure their weapons were available and ready.

This accounting of food, tools, armaments, and other commodities that were used by the community led to the creation of early numbering systems. These numbering systems eventually led to culture-based (community-based) written languages. The weaker men had a lot more time on their hands than the men who hunted and protected the community. With this abundance of time, these men perfected and developed their skills. Soon they learned to use these skills to their own advantage.

These men learned to entertain the children of the community while their mothers attended to those needs of the community that the stronger males did not attend to. The physically weaker men invented games for the children to play and then noticed *how* the children played. They were aware of the abundance of imagination that children use when they play. Seeing this, these ancient male babysitters began to invent stories that fascinated and entertained the children. Their babysitting techniques became very valuable to the mothers and fathers of the community, who had to work all day.

As the stories that the male babysitters told became more complex, language developed faster. In other words, the better their stories became and the more imagination they used, the more words they had to invent. The stories became more detailed as the children asked more questions.

Children are inquisitive and ask *a lot* of questions. Many innocent questions cannot be answered by their parents' limited knowledge. So if these male babysitters didn't know how to answer a child's question, they simply made something up. Children believe whatever adults tell them, especially the adults in whose care and trust their own parents have placed them.

At first, the stories the babysitters told were very simple and based on the people's limited knowledge. These ancient people didn't know much about the world in which they lived. They were too busy trying to survive and protect themselves from wild animals and other tribes in their limited sphere of existence.

If a child asked a question about the sun, for example, a story was invented to explain it. Fascinated and in complete awe and attention to the explanation, the child's mind began to learn things that were not real, taught to them by the make-believe stories.

When the parents got home from work and saw their children well cared for and happy, they wondered about the stories that created such excitement and entertainment for their children. At first, the parents realized that the stories were not real, but were just used to entertain their children. The children heard the stories time and time again over the years. Later, when the children grew up and became parents themselves, the stories they had been told as children became more real to them and began to affect their reality.

The aged children (now adults) who had heard the tales of fantasy from their babysitters began to desire *more* stories. The more the adults desired stories, the more complex and adult-like the stories became. After a long day's work, nothing was more relaxing than sitting around the fire listening to a good story.

The storytellers began to realize how important the stories were to the community. These stories became the oral history of the ancient world. Eventually, the storytellers developed a system of writing that helped them tell the same story over and over again.

The stories that benefited the unattractive men the most were used the most. Before long, the storytellers were just as important to the community as the hunters and protectors. In many cases, much more important.

Children ask questions that often cannot be answered by adults. Aged children (in other word, adults) also ask questions that cannot be answered by other adults. For this reason, some people make up answers to adult questions and then convince others that their answer is the correct one. When this happens, other adults find comfort in believing that at least someone (supposedly) knows the answers.

The invention of the concept of an all-knowing God gives comfort to people who do not know the answers. When the answer to the question is, "Well, God has all the answers," then the advantage goes to the person who can answer the question, "Who is God?"

The idea of "God" changes according to the knowledge of the people. If the people know the answers to their questions, then they don't need God, nor do they need the person who taught them about God.

The most important part of the stories told to people is the concept that there is a God who knows everything. To maintain their advantage over the masses, the storytellers had to convince the people of the existence of a God that only talked to them. This was because it increased their value in the community.

To gain advantage over their physically more attractive brothers, these accountants, these storytellers also began to use their power and position of trust to get sex from women. They held the keys to the community storage. They accounted for what was in that storage. And if a woman needed something from the storage while her mate was away hunting or protecting, the less attractive man could provide it, with a condition. These early accountants began to understand the power that they held over women.

But it didn't take long for the stronger men, who could not read, write, or tell stories, to realize what their weaker peers were doing. To the stronger men, the power and importance of the relatively inactive accountants, storytellers and shepherds (who sat around all day watching over the flocks), seemed unfair. In their jealousy or rage, caused by finding out that the community accountant had had sex with his woman, the stronger males began to kill the weaker men.

The weaker men had to do something about this physical disadvantage. At that time, they didn't have guns that would have given them the advantage over a stronger male. But they did have their stories.

Here is an example of one of the stories that was invented to keep the stronger males in check. We're going to tell it as it is told in modern times:

> And Adam knew Eve his wife; and she conceived, and bare Cain, and said, I have gotten a man from the Lord. And she again bare

his brother Abel. And Abel was a keeper of sheep, but Cain was a tiller of the ground. And in process of time it came to pass, that Cain brought of the fruit of the ground an offering unto the Lord. And Abel, he also brought of the firstlings of his flock and of the fat thereof. And the Lord had respect unto Abel and to his offering. But unto Cain and to his offering he had not respect. And Cain was very wroth, and his countenance fell. And the Lord said unto Cain, Why art thou wroth? [Why are you angry?] and why is thy countenance fallen? If thou doest well, shalt thou not be accepted? and if thou doest not well, sin lieth at the door. **And whatever happens unto thee, shall be your brother's desire, and he shall rule over thee.** [*Note the slight change to this last part. This part was correctly retranslated into English from its original Greek.*] And Cain talked with Abel his brother: and it came to pass, when they were in the field, that Cain rose up against Abel his brother, and slew him. And the Lord said unto Cain, Where is Abel thy brother? And he said, I know not: Am I my brother's keeper? And he said, What hast thou done? the voice of thy brother's blood crieth unto me from the ground. And now art thou cursed from the earth, which hath opened her mouth to receive thy brother's blood from thy hand; When thou tillest the ground, it shall not henceforth yield unto thee her strength; a fugitive and a vagabond shalt thou be in the earth.

In the story, Cain was physically stronger than Abel because of his constant toil in the fields while Abel sat around all day watching the sheep. This story explained what would happen to a stronger male if he killed a weaker one. The ancient storytellers/accountants who invented this story would also go on to invent many other stories that supported their role in the community.

Another example is the story that was told about a man named Isaac. He was deceived by his wife and his weaker son, Jacob (whose name was eventually changed to Israel). They tricked him into giving his blessing and birthright to the weaker son instead of to the physically stronger son, Esau, who rightfully deserved it. "God" sanctioned this deception in support of the weaker, but smarter, male.

These weaker men soon became the authority of the people. Their stories convinced the people to give them the "firstfruits of [their] labors," the best part of the community's produce, and the best parts of the people's flocks. In other words, these became the religious priests.

No one questioned their authority because no one questioned their stories. From the time that they were little children, the people had learned all that they knew—outside of the rudimentary daily tasks that provided them with their basic necessities—from the telling of stories. The children learned how to "till the ground" from their parents; but they learned how to please an imaginary god from the stories that they heard from their teachers. These were teachers that their parents were convinced knew a lot more than they did.

Remember, children only know what their parents teach them. Parents only know what their parents taught them. And outside of daily labors that support and protect a community, the only other knowledge that was available to ancient peoples came from the stories that they heard during childhood.

Many, many people continue to be influenced by these same stories today. Its effect on the mortal experience is devastating. The ancient philosophers (who had more time on their hands to sit around thinking) created these philosophies and religions to justify their existence and create value for themselves. As previously stated, human nature is responsible for these philosophies and the tragedies that they have created.

Chapter 5

True Messengers

If there was a religion that existed, that benefited humankind and created a more peaceful, equal, and caring world, then we would support it. But there is not.

We (the Real Illuminati™) understand how important spiritual beliefs are to the people living on Earth. Humans are driven to understand *why* they live upon Earth. Even more important to each of us, *everyone* wants to know, "What is my purpose for living?"

From our experience, we know that humans need religion as much as they need air to breathe and food to eat. Religion provides a person with food for the soul because it satisfies the human longing to understand one's purpose in life. Mortals do not know that each of us is an independent, highly advanced human being, connected to the Earth experience as a way of enhancing our otherwise perfect eternal existence. This is the cause of the God-presence, a feeling that we are not alone and that someone greater than our mortal Self is aware of us. It's this actual energy, this physical connection, that cannot be denied and that causes mortals to seek to understand what it is. When they can't find the answers themselves, they seek for someone who can provide them.

If knowing these things was possible for a normal person, then throughout history—with all the thinking and learning going on—somehow, somewhere, someone, would have come up with the right answers. But no one ever did, at least not from all the knowledge ever learned by a normal-functioning mortal brain.

Every religion attempts to answer these important questions. Unfortunately, false messengers try, but they can't. As we explained earlier:

When the answers do not come from within oneself, the sincere seeker looks elsewhere. No matter how sincere and humble the person is (recognizing their inability to find the answers), they are forced to look for someone outside of their own consciousness to bring them the answers: "I am looking for messengers."

Once open to the possibility that another human just might have the answers, the seeker of truth is presented with various ministers of knowledge, who provide answers made up of the "philosophies of men, mingled with scripture."

Unfortunately, most fall under the spell of false messengers who don't have any better understanding of the real truth than the supplicant who pays them for an understanding. False messengers' pay comes in a wide variety of forms. Most receive actual money or other relevant material goods for their services. But ALL receive a payment that is more enticing than all the money in the world: power over another person, respect and standing, and accolade-laced value from their followers.

Our goals have always centered on fighting these purveyors of injustice (these people who promote inequality and unfairness). We do not dominate them, but use them for our own purpose. When we couldn't find any of these merchants of knowledge (to whom the people looked to for the answers) who were worthy of our cause, we made our own messenger.

Throughout history, we have selected a few to represent us in our cause. We call this representative a "True Messenger." Whenever we saw the need, we asked a True Messenger to introduce new concepts of belief to the people among whom this messenger lived. It was our hope that upon so doing, the people might begin to support a religious movement that was good for humanity.

It's important here to understand that we cannot try to set up the right form of religion unless we have a person who is, not only willing to help us, but will sacrifice a normal life in the process. As we explained in a previous chapter, unless a person has experienced a brain *transfiguration* like all the other members of the Real Illuminati™, we will not reveal ourselves to him or her. Once the right circumstances exist where a new form of religious thought is possible, we need a True Messenger to help us do our best to influence the right form.

Free will is the most important part of being alive and living upon Earth. No one can be forced to think in any certain way. A person must choose, of their own free will, to think whatever it is that the person wants to think.

Certain events happen throughout the world in different cultures and communities. When we find a culture where the people have opened up their minds sufficiently, usually from the societal circumstances and current happenings of that particular time and place, we intercede behind the scenes. We travel throughout the world looking for such times and places. But even if we (the Real Illuminati™) find the right situation, without a person who is willing to become our True Messenger, we cannot proceed in our attempts.

Sometimes people began to think outside of the limited box of traditional prejudices that they had lived in for many years. Once the Greek civilization had developed enough, we decided to travel to that area of the world to see what we could do.

Near Athens, Greece, we found the man who the world would come to know as the ancient philosopher, Socrates. Socrates was the first to teach his students that those who tell the stories, also rule the world. We recruited him to be one of our True Messengers.

At that particular time, it was not our intent to set up any type of religion. It was our hope during the time of Socrates that we could influence people to think outside of the stories and traditions that gave "the few," power over "the many." What we asked Socrates to do was outside of religious belief.

Our intent has always been to get people to use their own brain, their own intuition, according to their own free will, to understand whatever it is that their *True Self* needs them to know.

Keep in mind, that each mortal is a *dream reflection* of their *True Self*. If it is not in your *True Self's* best interests that you know certain things while living a mortal existence on Earth, then no one has the right to force you to know something you shouldn't. Why? Because if you discover something that your *True Self* did not want you to discover, the discovery might mess up the way your *True Self* wants your mortal life to play out on Earth.

The problem with religion has always been that other people are telling you what you should do, how you should do it, what you should think about, and what conclusions you should make from your life experiences.

Through Socrates, we were able to introduce a way that we hoped would get people to discover things about themselves that would benefit their personal desires. Socrates did not tell anyone what they should do, how they should do it, what they should think, or what conclusions they should make about their personal life.

Each individual is different. Each individual has his or her own different needs. By asking each person *personal* questions, and providing the appropriate answers to their personal questions, it was our desire to stimulate a person's brain to answer their own questions. This pedagogy, this method of teaching, would become known as the *Socratic Method.* If this method had been successful, there would have been no need for us to try to get people to think on their own by using religion.

We had little success in Greek society at that time. The power of false messengers—who claim to have the answers that fit for everyone (i.e., their religious leaders)—was too overwhelming for the average person. Our efforts through Socrates failed.

We had cautioned Socrates about leaving any teachings in written form, knowing that if something is written, then it can be rewritten. According to our counsel, Socrates left no writings.

Unfortunately, some of Socrates' students misinterpreted his teachings in the presentation of their own. His few students found that they received value, prestige, and money in the community when they presented information that was politically and socially correct. So, they simply took some of

Socrates' information that wasn't politically and socially correct and made it so. Whatever they learned from our True Messenger, they changed it to conform to what their community wanted to hear, for which they were compensated handsomely (received a very good payment).

A few hundred years after Socrates lived, we again had the opportunity to introduce a new way of thinking. This "new way" countered the traditions and stories that the few used to control the many.

Once the Romans had assimilated the Greeks into their own community (first a Republic, then an Empire), freedom of thought among the masses increased.

Ancient stories had developed into religions of every kind. The Greek stories were precursors of the Roman stories. The Greeks developed their religions and the Romans developed theirs. These religions developed among the upper classes of economic division.

Greek and Roman gods favored the worldly successful. This happened by convincing the lower classes that the means of worldly success was not *earned* through hard work, but *given* as a "blessing" by the gods. This was very advantageous to the few in the upper class. Some stories were told about the gods choosing certain individuals as their mortal representatives. This helped the few (especially the physically weaker men) control the masses of the lower classes.

Some Greek stories were told by the weaker males. These stories always included the idea that the gods would choose a man and make him physically stronger than any other man. The people were convinced that if a person (male or female) was chosen by the gods, they would be given supernatural power, even if they were small in stature.

This concept: of god choosing a weak mortal and giving him or her supernatural powers—had worked well for the ancient Egyptian and Sumerian leaders. The Romans learned from these past successes. The religious leaders could appoint a *child* to be the choice of the gods, and the people would accept it. A child was much easier to control by the powers behind the scenes. What the people failed to see was how the child was being used by the real powers (the weaker men) who wanted the support and protection of the people.

Knowing how important religion is to a person who is searching for the meaning of life, and knowing how opportunistic people take advantage of this, we did everything we could to set up the perfect religion upon Earth.

Around 600 BCE, we wrote a religiously based dissertation (scripture) to confront what the ancient Hebrew culture had presented to their people about their culture's history. We asked our True Messenger to introduce it. Modern religious documents call him "Isaiah." Modern scholars however, have many questions about the actual existence of Isaiah. Where did Isaiah actually come from? Who was his father, Amoz? And why wasn't Isaiah's name associated with "the priests" like other major prophets were? (See Jeremiah 1:1 and Ezekiel 1:3.)

There is much more information included in Jewish history about the other prophets' interaction with the people than there is about Isaiah. Jewish historians were reluctant to include too much information about Isaiah. And once the real truth is known about him, the reason behind Jewish reluctance in not providing more information will begin to make sense.

Our messenger's true name was not Isaiah. It was Belzarach. His father's birth name was not Amoz. To make the text of his prophecies more compatible to and supportive of the Jewish faith, editors changed Belzarach's name to "Isaiah," which interpreted means, "God's salvation." They assigned the name "Amoz" (or Amos) to his father, which interpreted means, "burdened" or "troubled."

As we uncover the true facts about Isaiah, you will discover the Real Truth and finally understand why literary license was freely used to make these changes and present them as a factual historical account. Simply put, Jewish historians were not very proud of Belzarach's father and were embarrassed by the way "Isaiah" was treated by the leaders and the people of the kingdom of Judah.

In the scripture that we wrote for Belzarach, we called out this secret combination of political, philosophical, and business leaders who didn't think anyone was watching them ("their works were in the dark"). We called our movement "a marvelous work and a wonder." We explained that religious people were claiming to be good and honorable in the way that they were worshipping and honoring their god, but were actually hypocrites in the way that they acted towards each other.

Like religion often does, people say good things "with their mouth, and with their lips," but their actions are not good because "their fear [of God] is taught by the precepts of men," and they do what their religious leaders tell them to do.

What we wrote then has been translated into the English language in modern times:

Wherefore the Lord said, Forasmuch as this people draw near me with their mouth, and with their lips do honour me, but have removed their heart far from me, and their fear toward me is taught by the precept of men: Therefore, behold, I will proceed to do a marvellous work among this people, even a marvellous work and a wonder: for the wisdom of their wise men shall perish, and the understanding of their prudent men shall be hid.

Woe unto them that seek deep to hide their counsel from the Lord, and their works are in the dark, and they say, Who seeth us? And who knoweth us? Surely your turning of things upside down shall be esteemed as the potter's clay: for shall the work say of him that made it, He made me not? or shall the thing framed say of him that framed it, He had no understanding?

Our work, a Marvelous Work and a Wonder®, is to bring to light those things that the combination of political, religious, and economic powers do in the dark. Once we reveal what they have done throughout history and what they do today, these purveyors of injustice (secret combinations) will no longer be able to say, "It wasn't me who did that."

Perhaps false messengers might say that they didn't understand that what they were doing was bad for humanity: "Shall the thing say of him that framed it, He had no understanding?" Our intent is to make sure that the politicians, religious leaders, and economic leaders understand perfectly what they have "framed." And what they have framed as the places of worship where the masses go to hear God (to hear their religion) is their greatest machination (deception or scheme).

We wrote the book of Isaiah to counter what was written in the *Pentateuch* (the first five books of Moses). *(Note: the ancient Hebrew tribal leaders commissioned Greek writers to take their oral histories and transpose them into a written record. This is where the Pentateuch originated and the reason that it was first written in Greek.)*

If read correctly and honestly, the book of Isaiah promotes the idea that all religion is corrupt, and that a person should reject men as their authority: "Cease ye from man, whose breath is in his nostrils: for wherein is he to be accounted of?"

The original book of Isaiah was our attempt to persuade people to look <u>only</u> to God for guidance. In other words, we were trying to get people to think for themselves. At the time, people were convinced that there was a supernatural force outside of their own mind that would guide them, if they relied on it. Knowing that this force was actually a person's own mind, we presented the idea that a person should reject everything that they did not directly receive from God in their own mind.

While we had their attention, we explained the importance of taking care of the poor and needy, making this more important than any religious ritual or belief. (Consider reading Isaiah, chapter 1.)

Unfortunately, as is the case, anything that is written, can be rewritten. Later Jewish scribes interpolated many things into our book of Isaiah that put the "house of Israel" above all other "houses" and made the Jews their god's chosen people. This was not part of, nor intended by, our original writings for Isaiah. These parts were added later.

We knew how easy it was to change the written word, especially in ancient times when there were few copies and usually only one original manuscript. But we have the original manuscript of the book of Isaiah in our possession today. We also have a copy that we made of the original Greek Pentateuch. It is to these documents that we currently have our True Messenger refer in the presentation of our message to the modern world.

Chapter 6

Christianity

Many religions were set up for the rich. Nevertheless, there began to be a movement among the poorer people in the outer regions of the Roman Empire. This movement allowed us to recruit yet another True Messenger to our cause. We became aware of a child prodigy living in the Palestine region. His name was Inpendius.

From his birth, Inpendius' intelligence was equal to our own. He understood what we understood about human existence. Inpendius also understood things that we could not reveal to the public because of fear of being discovered and killed.

From a young age, Inpendius' savant-like intelligence impressed many people. At the age of twelve, Inpendius used modern CPR (Cardiopulmonary Resuscitation) techniques to revive one of his friends. This is a simple procedure according to modern knowledge, but no one had ever seen anyone performing CPR at that time. When his friend was revived, seemingly brought back to life from death, Inpendius' popularity among the masses began to grow.

His popularity was advantageous to our cause. Once of an age when other people in his culture would take him seriously, we moved to recruit Inpendius and make him our True Messenger.

Like Socrates before him, we counseled him not to leave any of his personal writings. We taught Inpendius how to influence others to think outside of the box of tradition in which the people were confined. We had learned from our experience with Socrates that it was better to utilize (use) the accepted traditions to influence individual free thought. Socrates did not do this. He did not use religion or "belief" in his teachings.

The people of Inpendius' day, living in his community, had successfully organized a religion that was based on their past oral histories. These had eventually been turned into written form by Greek writers as the *Pentateuch* (the five books of Moses in the Old Testament of the Bible).

Even among the poor masses, the few had power and control over the rest. The Hebrew priests personally benefited from the "laws of Moses," supposedly given to Moses by the one *true god*, according to the stories of the Bible.

Unlike Socrates, we helped Inpendius to spread our message by supplying him with our own ancient scripture: the book of Isaiah. We also gave him a few other writings that would eventually make their way into the canonized Old Testament.

We encouraged him to use the Socratic Method, but to ask questions and give answers within the religious beliefs of the people. By asking these relevant questions, Inpendius angered the false messengers. They accused him of corrupting the guidelines and instructions that God had given the people through them. In spite of counseling Inpendius to use the existing traditional religion in his efforts to open people's minds, he was killed (like Socrates before him) by those whose livelihoods his teachings threatened.

Although Inpendius confronted and confounded religion, he did not start one. His teachings were misinterpreted and misrepresented by his followers. When the stories about Jesus were written down, these misrepresentations overshadowed some of Inpendius' most important teachings. "The kingdom of God is within," for example, should have been understood as there is no god outside of one's own Self—the kingdom of this god dwelling within each person. This understanding did not support the power and control that religious leaders held over their followers.

After his death, Inpendius' popularity did not decrease (unlike what happens when most other people die). Stories about him were told among the poor. The god of the poor had made one of their own, into a hero. God gave the poor masses value in *not* choosing one of His principle mouthpieces from among the elite and entitled.

These unverified and unsubstantiated stories about Inpendius' life continued for hundreds of years among the poor communities of the Roman Empire. The Jewish leaders, who were wealthy and supported by the masses, argued against the idea that anyone outside of their priesthood lineage could be God's prophet. If God's messenger was poor, Jewish leaders would have a hard time justifying the way that they were supported by "God's will."

Eventually, the Great Roman Empire was threatened from within. The poor masses began to rebel more and more against the elite. Facing inevitable revolution and the threat of being overthrown, Roman authorities had to do something to save themselves and their power and control over the people.

The government secretly combined their influence with that of religious leaders. This secret combination led to the formation of a new religion that would cater to the interests and needs of the poor. The myths of Inpendius, the hero of long ago, became a source upon which the secret combination of religious and government power relied. They took advantage of the poor whom had embraced these stories. A new hero and religion were created: Jesus of Christianity.

The stories of Jesus were invented and used to quiet the masses, while still allowing government and religious leaders to prosper and take advantage of the poor. The teachings of Jesus spoke to the heart and soul of a believer. This was because they were cleverly centered on his teachings about the basic human principles of kindness, compassion, and, most importantly, tolerance of others.

The lower-class masses caused problems for the upper class of the powerful and elite. The economic upper classes persecuted the poor and despitefully used them in business to make a profit. The best thing for the rich would be to have the new Messiah command the people:

> Love your enemies, bless them that curse you, do good to them that hate you, and pray for them which despitefully use you, and persecute you.

In other words, do not rise up and revolt against the rich people. Subject yourselves to them and the way they treat you.

This new religion saved the Roman Empire from complete destruction and collapse. While the wealthy remained concentrated in Rome in the west, the masses were supporting the local leaders that gave them hope in a new religion forming in the east. Realizing the effect that the new religion had on the entire Roman Empire, Emperor Constantine moved the capital to the east.

The masses living in other parts of the failing empire were not on board at first with the new religion. The people of Corinth, Galatia, Ephesus, the Philippians, the Colossians, and the Thessalonians, as well as the people of Rome, took issue with the fact that the new god didn't acknowledge their existence and communities. To get them on board, the secret combination created another new hero: the apostle Paul, a missionary supposedly called personally by Jesus to take his message to them.

There is no Christian church on Earth today that has any of the original documents of the writings that they claim to be "God's word."

The so-named *Pauline Epistles*, which are reported as being the letters that the apostle Paul wrote, do not exist. They have never existed, because Paul never existed. Paul was a necessary invention to appease (console) the other Roman citizens who felt left out, because the new Roman religious hero, Jesus, was not from their particular communities.

We failed what we attempted through Inpendius during the rise of the Roman Empire. We used orthodox Jewish beliefs that were the strongest among the poor living in the area of Palestine. This is where we found Inpendius, whose mortal brain had been *transfigured* like ours (with the actual physical change taking place in the advanced Self's brain).

These efforts had led to the formation of Christianity, which spread throughout what remained of the fallen Roman Empire. This new religion allowed the controlling combination of politics and religion to subdue the minds and hearts of the masses. Their beliefs rallied people in support of church and state, which the people were convinced should be one and the same.

Although Christianity preached peace and equality, it did not practice it. Instead of supporting free will and individual happiness, Christians saw everyone who did not believe as they did as a threat. Christian leaders (both of the state and church) convinced the people that the Christian god was the only god to be worshipped and obeyed.

The people only had access to the Christian god through their Christian leaders (both church and state). This was advantageous to the leaders because they could determine what the Christian god wanted the people to do. No longer was "God's will" set in stone, as it literally was through the Jewish hero, Moses. Contemporary (modern-day) revelation through a living church and leader allowed full control of the people.

Christianity has lasted as long as it has because its early leaders replaced a strict, unchangeable God. He became one who changed and adjusted "His will" to the needs of the ever-changing and progressing cultures and their people. These continual changes have benefited church leaders, but have never been completely successful at creating peace on Earth by establishing a perfect society. This is mainly because the free will of the leaders often overrides the free will of the individual.

Having God communicate only through the writings of ancient prophets didn't need to be successful. Through His appointed contemporary leaders, God's relevant-day revelations would tell the people that nothing was going to change or get any better for them until He (the Christian god) came again to Earth to set up His kingdom for the last time. Until then, the people had their leaders to lead them and guide them. Of course, this doctrine was VERY advantageous to the leaders and the power and control that they held over the people.

Christianity did not prove to be the right type of religion that would bring peace and harmony to all of humanity.

Our True Messenger's (Inpendius') existence was hijacked and convoluted, not to deliver our message to the world, but to help the few continue their power and control over the rest.

What we tried through Inpendius (a.k.a. Jesus) didn't work. The people needed something more substantial that fit in with their religious beliefs. They needed direct communication to the God in which they believed. They needed the "word of God."

In the seventh (7th) century CE, nothing we attempted to do in the Roman culture worked to inspire individual free will and economic equality. Nothing we attempted to do in the newly created Christian culture seemed to work. We had our hand in actually influencing the construction and invention of the Jesus stories. Our greatest contribution was the teachings of Jesus in regards to what he was going to do when he came again: "When the Son of man shall come in his glory, and all the holy angels with him, then shall he sit upon the throne of his glory" to judge the world. (See Matthew, chapter 25.)

In that chapter, the "Son of man" makes no judgment on people who commit the sins that Christians believe will keep them out of heaven. The only judgment that is made by the "Son of man when he comes in his glory" (according to the story) is how well a person supported economic equality and equity throughout their life.

To dovetail (to fit together or work well with) this important teaching, we used the title, "Son of man," in our *Apocalypse* (our *disclosure*). No one ever made the connection. This is because the religious leaders would lose their power and control if the only reason for God's commandments and eventual judgments was about how well the human race established economic equality upon Earth. They needed people to think that God was going to punish them for not doing what their religious leaders said.

Further, because the teachings of Jesus were under the control of corrupt religious leaders, "God" gave them revelation and inspiration (literary license and authority) to change whatever part of Jesus' teachings was to their advantage, or to the advantage of the Church.

In the Jesus stories, where Jesus was asked what the greatest commandments were, the authors (influenced by our input) had Jesus originally respond,

> Love <u>thyself</u> with all thy heart, and with all thy soul, and with all thy mind. This is the first and great commandment. And the second is like unto it, Thou shalt love thy neighbor as thyself.

This created a problem for the Church. In loving your Self, you might create commandments and authorities that support your Self and this love for your Self. If you love your mortal Self more than a Church, you will follow your *own* rules and instructions instead of doing what a Church commands you to do. So why would you need a Church?

In order to control the people, who were convinced that their religious leaders spoke for God, a simple change of the word "thyself" to "God"—solved the problem. It ensured that individual free will and self-love remained in check and under the control of the false god that controlled their free will.

As we have mentioned, anything that is written, especially in ancient times, can be rewritten any way that the rewriters want it to be rewritten. The only source of proof would be the original documents. We have in our possession a handwritten copy of the original stories and teachings of Jesus, as they were first written in Greek.

Once Christianity had been established throughout Europe, we used our influence and knowledge to get our message out. We knew that we had to somehow convince the Christian religious powers that our message supported theirs. Our message has always been about the equality of the human race. We know that in order for equality to exist, first, economic equality has to be established.

To deliver our message in a form that appeared to support the new Christian message, we wrote the esoteric (cryptic) book of Revelation. We wrote Revelation in such a way that the religious leaders could not decipher our message as being anything other than one that agreed with and supported theirs. We were successful at getting Revelation into the *catholic* (meaning "worldwide" in Latin) canon of Christian scripture, which eventually became the New Testament of the Bible.

We used ideas and concepts found in the Old Testament as symbolic representations that would eventually deliver our message to the world, but only if deciphered and explained by one of our own—by a True Messenger.

The Greek word for "disclosure" is "*apocalypse.*" This is what we originally called our book. "Revelation" serves the idea of disclosure well. When read <u>correctly and in context</u>, the book of Revelation does not foretell of the end of the world. It speaks of "renewing" the Earth. There is no mention of a worldwide *Apocalypse* in the text of the book, which corrupt religious leaders describe as the end of the world. They describe the symbolism in Revelation that they cannot properly explain in a way that scares their followers into submission to the "will of God."

The *disclosure's* message is simple and straightforward. We present a hero riding on a white horse, from whose mouth a sharp two-edged sword comes. We invented this figurative hero by borrowing from Greek mythology. Perseus was a mortal-god chosen to help protect mortals by fighting against evil. When Perseus finally defeated the woman Medusa, a white, winged horse leapt out of Medusa's blood. We used a woman figure similar to Medusa in our figurative presentation to represent the evils of the world:

> And upon her forehead [Medusa's] was a name written, MYSTERY, BABYLON THE GREAT, THE MOTHER OF HARLOTS AND ABOMINATION OF THE EARTH.

Our *disclosure's* hero would come to the world's rescue and destroy the woman with the sword that came out of his mouth. Our Medusa was economic inequality caused by the "kings" (leaders of both church and state) and the "merchants of the earth."

In our *apocalypse*, <u>only</u> the political, religious, and business powers on Earth are affected by our *Perseus-*

like hero riding in on a white horse and cutting off the head of the woman upon whose forehead the *real* sins of the world are written. There is no mention of the unbelievers (non-Christians), or those who violate Christian authority and commandments, suffering in the least:

> And the kings of the earth, who have committed fornication and lived deliciously with her, shall bewail her, and lament for her, when they shall see the smoke of her burning, Standing afar off for the fear of her torment, saying, Alas, alas, that great city Babylon, that mighty city! for in one hour is thy judgment come. And the merchants of the earth shall weep and mourn over her; for no man buyeth their merchandise any more[.]

The mythical Greek beauty, Medusa, "fornicates" with men who "lived deliciously with her," until what was actually on her head, coming out of her forehead, destroys them.

Nothing is more important to individual free will than first, the ability to exercise this will according to the happiness of the person, and second, the economic security needed to do so. Before the first can be realized, the second must be secured.

We have written books that were successfully introduced into the religious structures of this world. All of these books speak of economic equality and equity. The book of Isaiah speaks of it and the book of Revelation profoundly drives the message home.

Chapter 7

Islam

As Christianity spread, its father, Judaism, didn't so much. Judaism began to lose traction as Christianity grew. In their fight for value, Jews argued that Christianity wouldn't exist without Judaism and its history. There wouldn't be a New Testament unless an Old Testament existed. They had a point.

The Jews began to point out how corrupt the Christians were becoming. At that time, the Jewish Church began to have some success among the poorer cultures of Christians that were developing in that part of the world (Israel).

Regardless of ethnic origins or religion, the family unit became the basis and most important institution among the developing Christian faith. The Jews already had established the importance of the family, and they were backed up by the scriptures that both religious groups believed were the "word of God." The Bible specifically outlined God's chosen people—from whose lineage the Messiah of all humanity would come—as the family of Abraham, Isaac, and Jacob (the house of Israel). The Bible said nothing about the other families that existed in the world at that time. If you weren't from Jewish lineage, you weren't part of God's chosen people.

Economically, at that time (400 CE to 600 CE), the Christians were much more prosperous than the Jews were. This was because they belonged to and supported the flourishing Christian system of faith and commerce spreading throughout the post-Roman world. The Christians argued that they were receiving more of God's blessings than the Jews were. Therefore, how could it be possible that God didn't appreciate them just as much as His "chosen people"? These beliefs caused many divisions and arguments among the people.

This is when we made another effort to introduce our message to the people. We had hope of initiating a change of mindset to help people to work together to create the perfect society instead of fighting with each other.

In the area where the city of Mecca developed, there was a certain family that prided itself in its economic success in business and in its political popularity in the community. This family was known as the Quraysh.

A man named Hashim ibn Abd-Manaf was Quraysh. (The men didn't have last names. They were known as the "son of" [ibn]. Hashim was the son of Abd-Manaf.) Hashim wasn't as successful as some of his other brothers, and lived (in comparison to his brothers) poor. According to the traditional pride of the Quraysh family, if you were Quraysh and poor, you must be doing something wrong that was displeasing God.

Hashim had a son named Abd al-Muttalib. Now, you might wonder why Abd wasn't named *Abd ibn Hashim*, after his father, as was the custom at the time. Before Abd was born, his father Hashim was murdered by one of his cousins for trying to defend his honor in the family. Abd was raised by his uncle, Muttalib, who was also poor. His uncle told Abd what had happened to his dad, which was a very sore spot for Abd, who began to resent his Quraysh heritage.

Abd had a son named Abdullah, to whom he passed on his disgust of the inequality in the Quraysh family. Abdullah had a son named Muhammad.

Muhammad was a good kid and had a hard time listening to all the bickering going on with the members of his family. They had now divided even further, not only along economic lines, but also because of religion and ideologies.

The Quraysh family members had many different beliefs at that time. Some were Christians, some were Jews, some were Zoroastrians, and others belonged to other different religious sects of that day. Regardless of what a member of the Quraysh believed, if you were poor, you couldn't be in the right religion.

As Muhammad grew up, he heard it all. Muhammad's family was poor. Muhammad didn't have the chance of receiving formal schooling. By the time he reached age 25, Muhammad was completely confused about religion and truth. In his 25th year (actually 600 CE), in the month known on the Muslim calendar as Ramadan, while wandering alone on the outskirts of Mecca, as he often did, Muhammad cried with all of his might to a god that he didn't know nor understand.

"Are you a Christian? Are you a Jew? Who are you God?" he cried in anguish. An instant feeling of intense energy filled his body. From that moment, Muhammad knew that none of the religions were correct.

At that time, Muhammad's physical brain had actually experienced what we have termed a *transfiguration*. His brain was transfigured in such a way that he could understand things about human existence that few other mortals could. This actual, physical *transfiguration* had happened similarly to Socrates. But again, it had been different for Inpendius. He had been *born with* the knowledge, and was seen as an anomaly, a savant, or a child prodigy.

After his *transfiguration*, Muhammad began to explain ideas that no one else had ever considered. He would say things that made a lot of sense, but that countered many of the traditions and beliefs that his family and friends held dear.

He started to contend with many of the respected elders in the Quraysh tribe, including those who were Christian, Jew, Zoroastrian, and others. His words confounded them, which they didn't take too lightly. Muhammad's experiences, after his mind was opened to Real Truth and as he began to tell others about what he knew, were exactly what a later True Messenger would report about his own:

> I soon found, however, that my telling the story had excited a great deal of prejudice against me among professors of religion, and was the cause of great persecution, which continued to increase; and though I was an obscure boy, only between fourteen and fifteen years of age, and my circumstances in life such as to make a boy of no consequence in the world, yet men of high standing would take notice sufficient to excite the public mind against me, and create a bitter persecution; and this was common among all the sects—all united to persecute me.

> It caused me serious reflection then, and often has since, how very strange it was that an obscure boy, of a little over fourteen years of age, and one, too, who was doomed to the necessity of obtaining a scanty maintenance by his daily labor, should be thought a character of sufficient importance to attract the attention of

the great ones of the most popular sects of the day, and in a manner to create in them a spirit of the most bitter persecution and reviling. But strange or not, so it was, and it was often the cause of great sorrow to myself.

One of us (the Real Illuminati™) heard of Muhammad and began to follow him and listen to the things that he discussed openly. This one reported Muhammad's enlightenment to the rest of us. We watched Muhammad for many years. We wanted to see how he handled what he now knew: things that had been hidden since the foundation of the world, and things that few other mortals understood. Ten years later, we first approached Muhammad and revealed ourselves to him. He was about 35 years old at that time.

During all those years after his *transfiguration*, no one understood the things that Muhammad did. He learned not to discuss them with others, which had led to a very lonely existence. As he was accustomed to do, Muhammad spent a lot of his free time by himself. He wandered in the country on the outskirts of Mecca, away from all the noise of the city where he lived.

We followed him to his favorite places, where we finally approached him. When we revealed that we too had experienced the same *transfiguration* that he had experienced, and knew what he knew, Muhammad was emotionally overjoyed. He had found others like him! We became his friends.

He loved being around us and would often ask us our opinions of life, including religion, politics, and economics. The more we shared with Muhammad about who we were and what we knew, the more excited he became to join our small group. We recruited him as our True Messenger.

More than from anyone else, Muhammad had received persecution and alienation from his own family (the Quraysh). Muhammad's immediate family was from the poor side of town. Other members of the Quraysh tribe were from the rich side. The wealthier, more educated members of the Quraysh family had mocked Muhammad for making things up, to make himself look more educated and important than they thought that he was.

We knew there was no way that Muhammad would be able to help us improve the lot of humanity by telling the Real Truth about human existence. Especially, there was no way that people would accept that there is no god outside of our own brains, and that we are solely responsible for everything that we do, good or bad, during our life upon Earth. We watched Muhammad try to introduce some facets of Real Truth to others, only to be rejected, alienated, and persecuted.

It was our intent to try something different with Muhammad. We wanted him to start a new religious movement based on the Real Truth, yet presented in a way that supported the orthodox religious beliefs of his day. We did not want him to write any scripture, and he never did. Again, anything that is written can be rewritten to reflect whatever the rewriter wants the reader to believe.

To present what we wanted humanity to know about the necessary changes needed to bring about a perfect society on Earth, we knew that the information had to come from the god in which the people already believed. People didn't want the *Real* Truth; they wanted information presented to them that fit in with and supported those things that gave them value and worth.

Much later, when we finally *did* write our own scripture in the early nineteenth (19th) century, we put it this way:

> But behold, the Jews [in this case the Arabs] were a stiffnecked people; and they despised the words of plainness [the Real Truth], and killed [and persecuted] the prophets [True Messengers], and sought for things that they could not understand. Wherefore, because of their blindness, which blindness came by looking beyond the mark [of plainness established by the Real Truth], they must needs fall; for God hath taken away his plainness from them, and delivered unto them many things which they cannot understand, because they desired it. And because they desired it God hath done it, that they may stumble.

Muhammad had spent ten years trying to get people to think. His experience had proven to him that there was no other way to get people to open their minds to something different than by using their *incorrect* traditions and beliefs. He realized that he had to utilize (use) the lies that they already accepted as truth.

The Old Testament stories about Adam as the first man, and the Hebrew prophets and kings being chosen by the Hebrew god, were a lie. Muhammad often debated the authenticity of the Jewish stories that led to Christianity. These debates only led to more persecution.

His experiences during the decade before we approached him gave Muhammad all the proof he needed. He agreed to help us help humanity the only way that people would allow: through accepted and believed lies—through religion. Therefore, we invented acceptable lies that would help people accept Muhammad's wisdom, as if it came from God. We did the same for the other True Messengers who came after Muhammad.

Our efforts through Muhammad led to the formation of Islam. After the development of Islam, from our discussions and interviews with him, we helped put together a brief history of Muhammad's early life and how he became one of the most important figures in world history.

The following was written by us about Muhammad. We wrote it with the intention of getting it included in later Muslim history, which we failed to do. We wrote it in Muhammad's own voice in order to reflect a personal undertone.

(Excerpt from a history of the early life of Muhammad):

Owing to the many reports which have been put in circulation by evil-disposed and designing persons, in relation to the rise and progress of the Islamic faith, all of which have been designed by the authors thereof to militate against Islam's character as a religion and its progress in the world—I have been induced to dictate this history, to disabuse the public mind, and put all inquirers after truth in possession of the facts, as they have transpired, in relation both to myself and the Islam religion, so far as I have such facts in my possession.

In this history I shall present the various events in relation to the beginnings of Islam, in truth and righteousness, as they have transpired, or as they at present exist, being now the eighteenth year since I was first called by God.

I was born in the year [575 CE], on the 2nd day of the recognized month of maymūn, in the city of Mecca, located in the southwest of Mesopotamia. I belonged to the Quraysh tribe. My father's name was Abdullah. Like his grandfather before him, my father was murdered by some of his cousins before I was born. Fearing for her life, my mother fled [to a more humble part of] the Quraysh tribe and lived with a kind family that belonged to another part of my ancestral line. My mother soon died thereafter, and this family was too poor and unable to care for me, so they returned me to my father's family. I was raised primarily by my uncle, Abu Talib.

Throughout the Arab world, at the time of my youth, and more particularly in Mecca, there was an unusual excitement on the subject of religion. It commenced with the Christians and the Jews, but soon became general among all the sects in that region of country. Indeed, the whole district of country seemed affected by it, and great multitudes united themselves to the different religious parties, which created no small stir and division amongst the people, some crying, "Lo here!" and others, "Lo, there!" Some were contending for the Christian faith, some for the Jewish, and some for the Zoroastrian.

For, notwithstanding the great love which the converts to these different faiths expressed at the time of their conversion, and the great zeal manifested by the respective clergy, who were active in getting up and promoting this extraordinary scene of religious feeling, in order to have everybody converted, as they were pleased to call it, let them join what sect they pleased; yet when the converts began to file off, some to one party and some to another, it was seen that the seemingly good feelings of both the priests and the converts were more pretended than real; for a scene of great confusion and bad feeling ensued—priest contending against priest, and convert against convert; so that all their good feelings one for another, if they ever had any, were entirely lost in a strife of words and a contest about opinions.

I was at this time in my twenty-fifth year. The different families of the Quraysh were greatly divided on the subject of religion. Both my great-grandfather and my father were killed in religious disputes. This greatly affected my mind and caused me much anxiety when I pondered on the subject of religion.

During this time of great excitement my mind was called up to serious reflection and great uneasiness; but though my feelings were deep and often poignant, still I kept myself aloof from all these parties, though I attended their several meetings as often as occasion would permit. In process of time my mind became somewhat partial to the Christian faith, and I felt some desire to be united with them; but so great were the confusion and strife among the different Christian denominations, that it was impossible for a person to come to any certain conclusion who was right and who was wrong.

My mind at times was greatly excited, the cry and tumult were so great and incessant. The Christians were most decided against the Jews, and used all the powers of both reason and sophistry to prove their errors, or, at least, to make the people think they were in error. On the other hand, the Zoroastrians in their turn were equally zealous in endeavoring to establish their own tenets and disprove all others.

In the midst of this war of words and tumult of opinions, I often said to myself: What is to be done? Who of all these parties are right; or, are they all wrong together? If any one of them be right, which is it, and how shall I know it?"

(End of excerpt.)

Muhammad could not read or write. There was no way he could have written his own history. Many years after he became a spiritual leader (a prophet who people think speaks with God), we wrote this brief history for him.

We knew Muhammad long before he became a prophet in the eyes of the people. We knew, then, of his lack of education and inability to read and write. It was our hope that one day he would learn. If he *had* learned, we would have been able to introduce the brief, honest history about his early days (given above), written by his own hand. But Muhammad never learned to read or write.

Hundreds of years later, we used what we wrote for Muhammad as a blueprint for the same brief history of the early years of a young man we chose as our True Messenger in the United States of America.

Although very uneducated like Muhammad, this young man learned to be a proficient writer, but just not at first. Many years after his followers organized a new American religion, we instructed him to write his own brief history. We had him base his personal history on the format we had used to write for Muhammad, hundreds of years before.

The fact that Muhammad was illiterate is evidence for his followers that the Quran might not have been written by Muhammad. If Muhammad wasn't literate enough to write the Quran, then who did? It's easy to accept that the book was written by later followers (later scribes), who wrote down what they claimed Muhammad had said. But even more important to the Islamic faith is, "Where did the things that Muhammad said come from?"

Orthodox (accepted) history got many things wrong not only about Muhammad's early life, but also about the beginnings of Islam. (We have presented the exact day of his birth above, which is a date disputed by many scholars.) What Muhammad actually said and did, and what was reported by others about what he said and did, are sometimes very different. There were no recording devices at that time; and much of the time when Muhammad spoke, no one had the proper writing tools to make an exact record.

The Quran is a collection of later writings that a few of Muhammad's closest companions wrote. This was their attempt to pass on what they claimed they had heard from his mouth. In fact, it was almost 100 years later, at the height of the success of Islam throughout the Middle East (Mesopotamia), that Muslim leaders finally canonized what would become the Holy Quran.

Baghdad became the Islamic capital in the eighth (8th) century CE. It became a center of great learning and study. In Baghdad, the Muslim Caliph (head prophet) organized a group of the best Muslim writers that he could hire to create an official compilation of Muhammad's teachings. Before this time, there were many different writings that were being passed among the different Islamic tribes (families), which were reportedly written by Muhammad's close companions.

History had once again repeated itself. The Inpendius (Jesus) stories and teachings were reported as being written by Jesus' close companions—firsthand witnesses. The exact same thing happened to Muhammad.

A compilation of Muhammad's teachings led to an orthodox (accepted) and official catalogue: the Holy Quran. The original collection of his teachings was supposedly written by the disciples who had accompanied him throughout his life. It was created exactly like the canonization (turning it into "scripture") of the teachings of Inpendius. These supposed stories and lessons from the "teacher of righteousness" mentioned in the modern Dead Sea Scrolls became the Christian New Testament.

Neither Inpendius nor Muhammad left any personal writings, just as we had counseled Socrates not to leave any. Instead, the teachings of our True Messengers were later rewritten and compiled to be used by unscrupulous and opportunistic men to gain power and control over their followers.

We have yet to reveal the exact details of how we found Socrates and Inpendius and recruited each to be our public messenger. We have avoided these details because of the lack of accepted historical documentation that would substantiate that either of these men actually existed as orthodox history reports.

The Messengers that we chose later however, beginning with Muhammad, have enough historical documentation written about them to allow an honest researcher of *Real* Truth to consider what we are going to reveal in this book.

Chapter 8

A New American Scripture

The United States of America was established on the principle of freedom of religion. The U.S. Constitution was supposed to afford each citizen with the right and the protection to practice whatever form of religion one wanted. Once the United States was fully established in the Western Hemisphere, we found another chance to try to help the human race.

We first approached Thomas Jefferson. At the time, we could find no other person who had experienced a brain *transfiguration* like the members of our group had; but Jefferson appeared to be open-minded and an individual who fought for equality and justice for all.

Around this time period, secretive groups were being formed in both the United States and throughout Europe. Some of these groups evolved into what is known as Freemasonry. None of our members of the Real Illuminati™ was involved, in any way, in the formation of these groups. We involved ourselves with the poor majority masses. These groups involved themselves with the educated and rich few.

We were attracted first to Thomas Jefferson because of his public ridicule of these secretive groups. We presented ourselves as former members of the Masonic movement that had originated in Europe. Our knowledge of these secret groups, our knowledge of religion, and our knowledge of almost everything else impressed Jefferson.

We presented an idea to Jefferson that might help the newly founded United States become a good nation, a righteous nation, a nation of good people. Knowing that most early Americans were staunch Christians, we explained what our intent was to Jefferson. We asked for his help.

Jefferson's brain had not been *transfigured*. He did not know what we knew. He was also a very popular politician. He was not the right candidate for a True Messenger.

We explained to Jefferson that the core teachings of Jesus as presented in the New Testament were solid. We told him that they could be the basis for the right religious movement. We explained that the native American people had as many rights to life, liberty, and the pursuit of happiness as the European Christians, who now occupied their ancestral lands. We explained how we wanted to introduce a religious history of the native American people that would link them to the people described in the Bible, the ancient Hebrews.

Jefferson agreed that something needed to be done to bring more equality and justice to the native Americans. He was also concerned about the poor people of this new, growing and developing nation. But he disagreed that forming a new religious movement based on new, written scripture was the right way to do it.

Jefferson would not help us establish a new religious movement. Instead, he took our ideas and wrote his own form of scripture. He titled his book:

The Philosophy of Jesus of Nazareth, Extracted from the Account of His Life and Doctrines as Given by Matthew, Mark, Luke and John; Being an Abridgement of the New Testament for the Use of the Indians, Unembarrassed with [does not include] Matters of Fact or Faith beyond the Level of their Comprehensions.

Jefferson's Bible failed to convince the American people that the native Americans deserved the same rights afforded to the Americans by their Christian God. The white-skinned, European-American Christians had to somehow justify the taking of native American lands and killing the people. The justification: the natives didn't follow Christ and did not have any part of "God's chosen people," nor a right to the new American "Promised Land."

Shortly after Jefferson published his religious essay, we were made aware of the birth of a child to a poor American family living in the northeastern part of the United States. We sought out this family and watched the boy grow, wondering if it were possible that he had been born with a *transfigured* brain, as Inpendius had been. Upon observation, we discovered that he was a normal boy. Nevertheless, we continued to monitor his development the best we could.

We discovered a popular preacher by the name of Ethan Smith who was living in the same area of the United States where this child was living. Because he was a preacher and had a rather large following, we approached him. We began to influence him with ideas about the native American people actually being part of what the Christians believed were the Ten Lost Tribes of Israel.

The accepted Christian story from the Bible taught that God selected Abraham and his family as his "chosen people." Abraham's wife, Sarah, couldn't have kids, so she gave her handmaiden, Hagar, to Abraham so that Abraham could fulfill the promise God had made to him. Hagar was of a darker complexion than Sarah. Her son, Ishmael, was born.

Sarah was a bit jealous and wanted her own son. So God blessed Sarah in her old age and gave her a son, Isaac. From Isaac's lineage came the lighter-skinned Jews. From Ishmael's lineage came the slightly darker-skinned Arabs. To this day, the Jews and Arabs still fight over which group is God's chosen people through Abraham. They are fighting over a story that isn't even true.

Isaac had a son named Jacob. As the story goes, Jacob was a weak man, unlike his older brother Esau, who was a strong hunter and warrior. Of course, the Hebrew God chose the weaker Jacob as a leader over his stronger brother Esau. Through deception (which obviously the Hebrew God is okay with), Jacob stole the birthright from his older brother.

God eventually changed Jacob's name to Israel. Israel had twelve sons. Each of his sons had his own lineage. Jacob (Israel) had a few wives, but his favorite wife was Rachel. Rachel was prettier than her older sister, Leah. But according to tradition, Jacob had to take the uglier Leah to wife before he could have the younger, prettier one that he really wanted. Again, this is according to the stories in the Bible, none of which are true.

Two of Israel's sons, 1) Judah from Leah, and 2) Joseph from Rachel, had a bunch of kids. Their descendants later argued among themselves about who was the more chosen lineage. So Judah's kids set up the kingdom of Judah and Joseph's kids set up the kingdom of Israel, in honor of the specialness of being from Israel's favorite wife.

The question that many historians and scholars (the false messengers) argue about is, "What happened to the other ten brothers' descendants? The Bible provides a history of Judah's and Joseph's houses, but what about the whole 'house of Israel'? Where *are* those ten lost tribes?"

If the Bible story was true, then it could have explained what happened to these other ten sons of Israel. But those who created the Bible story didn't really care if they explained about anyone's history but their own. Because the Bible was written hundreds of years after Abraham, Isaac, and Jacob supposedly existed, and because it was written for and about the house of Judah (or the Jews), what happened to the other sons didn't matter much. But these Jewish false messengers left something undone in the reporting of their history. They actually mentioned the house of Joseph, through his sons, Ephraim and Manasseh.

It is said that the house of Ephraim became the Assyrians and the Sumerians. But who really knows? The Jews don't care, as long as they are known as the main, chosen "house of Israel." But where did the descendants of Joseph actually end up? This is a bit more difficult for any false messenger to explain.

Knowing what we know about human history, and also about religious history, we were able to capitalize on the holes left by the Jewish false narrative. To a lot of people in this world, the Jewish narrative found in the Bible is not false, but a true history of events. It couldn't be any more false. Still, this false narrative is accepted by so many people as truth and is the basis for three of the world's most popular religions: Judaism, Christianity, and Islam. Therefore, we knew that if we were going to write our own scripture in an effort to help create the *right* kind of religion, we would be forced to use what we knew was false.

In order to get people to read our new scripture, we formulated the storyline on what happened to the tribe of Joseph once Jerusalem was destroyed in about 600 BCE. When we met with Ethan Smith, we discussed the possibility (we made up the possibility) that God would reveal the true history of the native American people as being from the house of Israel, through Joseph (Israel's son). Instead, Ethan Smith plagiarized our idea (he stole it) and wrote his own book called *View of the Hebrews*.

A few years before Ethan Smith published his book, we learned that the young boy whose life we were monitoring had experienced the same brain *transfiguration* that we had. His name was Joseph Smith, Jr. (No relation to Ethan Smith.)

Joseph Smith had his *transfiguration* in the spring of 1820, when he was fourteen years old. We did not introduce ourselves to him at that time. We wanted to watch him mature and see how he would use the knowledge that one receives when their brain is more fully connected to their *True Self.*

It was our intent to wait until Joseph Smith was in his mid-twenties before asking him to join us. But the preacher Ethan Smith published his book, which caused quite a stir among the Christians in that area. We had to act fast. We had to counter Ethan Smith's false, plagiarized (stolen) information that we had discussed with him many years earlier. The same year that the *View of the Hebrews* was published, we were forced to reveal ourselves to the then 17-year-old teenager, Joseph Smith. Jr. We recruited Joseph as our True Messenger.

We asked Joseph to help us form the right type of religion that we felt would open the minds of the

American Christians of the early nineteenth (19th) century. We wrote the *Book of Mormon* as a source of "God's will" (scripture) to aid us in our attempt. This book was our final attempt at using religion to get people to think. We wrote it in such a way that it would be hard to dispute, according to the information that was given in the Bible.

In our new Americanized scripture, we reintroduced Isaiah as the most important book of the Bible. We also included mention of the book of Revelation and its profound importance.

In the storyline of the *Book of Mormon*, we finally introduced ourselves the way that we felt the world would be able to accept our existence. We provided clues to help people understand what we do behind the scenes. We explained our *transfiguration* and the overall intent of our work through the words of our main character, Mormon. Again, this is how we, the Real Illuminati™, introduced the concept of our existence. We did this through the story that we wrote for the *Book of Mormon*:

> And behold, the heavens were opened, and they were caught up into heaven, and saw and heard unspeakable things.
>
> And it was forbidden them that they should utter; neither was it given unto them power that they could utter the things which they saw and heard;
>
> And whether they were in the body or out of the body, they could not tell; for it did seem unto them like a *transfiguration* of them, that they were changed from this body of flesh into

an immortal state, that they could behold the things of God.

But it came to pass that they did again minister upon the face of the earth; nevertheless they did not minister of the things which they had heard and seen, because of the commandment which was given them in heaven.

And now, whether they were mortal or immortal, from the day of their *transfiguration*, I know not;

But this much I know, according to the record which hath been given—they did go forth upon the face of the land, and did minister unto all the people, uniting as many to the church as would believe in their preaching; baptizing them, and as many as were baptized did receive the Holy Ghost.

... And it came to pass that thus they did go forth among all the people of Nephi, and did preach the gospel of Christ unto all people upon the face of the land; and they were converted unto the Lord, and were united unto the church of Christ, and thus the people of that generation were blessed, according to the word of Jesus.

And now I, Mormon, make an end of speaking concerning these things for a time.

Behold, I was about to write the names of those who were never to taste of death, but the Lord forbade; therefore I write them not, for they are hid from the world.

But behold, I have seen them, and they have ministered unto me.

And behold they will be among the Gentiles, and the Gentiles shall know them not.

They will also be among the Jews, and the Jews shall know them not.

And it shall come to pass, when the Lord seeth fit in his wisdom that they shall minister unto all the scattered tribes of Israel, and unto all nations, kindreds, tongues and people, and shall bring out of them unto Jesus many souls, that their desire may be fulfilled, and also because of the convincing power of God which is in them.

And they are as the angels of God, and if they shall pray unto the Father in the name of Jesus they can show themselves unto whatsoever man it seemeth them good.

Therefore, great and marvelous works shall be wrought by them, before the great and coming day when all people must surely stand before the judgment-seat of Christ;

Yea even among the Gentiles shall there be a great and marvelous work wrought by them, before that judgment day.

And if ye had all the scriptures which give an account of all the marvelous works of Christ, ye would, according to the words of Christ, know that these things must surely come.

And wo be unto him that will not hearken unto the words of Jesus, and also to them whom he hath chosen and sent among them; for whoso receiveth not the words of Jesus and the words of those whom he hath sent receiveth not him; and therefore he will not receive them at the last day;

And it would be better for them if they had not been born. For do ye suppose that ye can get rid of the justice of an offended God, who hath been trampled under feet of men, that thereby salvation might come?

And now behold, as I spake concerning those whom the Lord hath chosen, yea, even three who were caught up into the heavens, that I knew not whether they were cleansed from mortality to immortality—

But behold, since I wrote, I have inquired of the Lord, and he hath made it manifest unto me that there must needs be a change wrought upon their bodies, or else it needs be that they must taste of death;

Therefore, that they might not taste of death there was a change wrought upon their bodies, that they might not suffer pain nor sorrow save it were for the sins of the world.

Now this change was not equal to that which shall take place at the last day; but there was a change wrought upon them, insomuch that Satan could have no power over them, that he could not tempt them; and they were sanctified in the flesh, that they were holy, and that the powers of the earth could not hold them.

And in this state they were to remain until the judgment day of Christ; and at that day they were to receive a greater change, and to be received into the kingdom of the Father to go no more out, but to dwell with God eternally in the heavens.

Although there are more than three of us, we are known by the LDS/Mormons as the *Three Nephites*, based on the way that we presented clues to our existence in our new scripture.

Upon writing this scripture for the American Christians, we had no idea if we would be successful at setting up a correct religion. This religion would need to satisfy a person's longing to know the answers to life's questions and yet also set up an organization that would promote peace and harmony on Earth. It wasn't long after we had Joseph Smith publish our book of scripture in 1830 that we found out that we had again failed through religion.

In our new scripture though, we provided a way that we could rectify (fix) any bad religion that might develop from the people misinterpreting or ignoring our message in the *Book of Mormon*. We introduced the existence of the book's *sealed portion*. This "sealed" part of the story was withheld from the American Christians to see what they would do with the new scripture and what kind of religion they would develop from it. We gave various clues about this throughout the new scripture:

> And these things have I written, which are a lesser part of the things which he taught the people; and I have written them to the intent that they may be brought again unto this people, from the Gentiles, according to the words which Jesus hath spoken.

> And when they shall have received this [the *Book of Mormon*], which is expedient that they should have first, to try their faith, and if it shall so be that they shall believe these things [the *Book of Mormon*] then shall the greater things be made manifest unto them.

> And if it so be that they will not believe these things, then shall the greater things be withheld from them, unto their condemnation.

The Mormon religion (the Church of Jesus Christ of Latter-day Saints) was the result. Not only do Mormons reject the idea that the "Three Nephites" are working behind the scenes, but they have entirely ignored our book's message and intent. They do not think that they are being condemned, when we made it perfectly clear that unless they have received the "greater things," they are condemned.

The LDS/Mormon Church has become one of the wealthiest religions in modern times. The numbers of their members, which reflect how many people accept and adhere to certain religious beliefs, is almost the exact same number as that of the modern Jews. We pondered (thought about) how we were going to write the storyline for our *Book of Mormon*. We decided to utilize (use) Jewish beliefs (the same beliefs). We knew that by doing this, probably the same number of Earth's inhabitants would embrace this "philosophy of men mingled with scripture." Both religious sects believe that their religion is the only true religion. Both religions believe that God will save the world through their religion and its leaders.

To counter the false messengers of Mormonism and the corrupt religion that has resulted from our *Book of Mormon*, we finally wrote the *sealed portion* of the book and published it through another of our recruited True Messengers.

Many Christians who read our *Book of Mormon* have been converted to the religion that shares it, only to be disappointed by the church that has developed. But their testimony of the feelings that they had when they read our book is not so easily denied.

Many people have joined the Mormons *because* of our book. Many people have left the Mormon Church because it does not adhere to the teachings that we promote in our book. The Mormons *use* our book, not to promote our teachings and intent, but to promote their own corrupt religion, based on the teachings of their false messengers.

The power of our writings is undeniable. If read by a sincere Bible-believing person, the truthfulness of the *Book of Mormon* will undoubtedly influence an acceptance of the book, just like the reader had accepted the Bible. Likewise, the proof that we wrote *The Sealed Portion* of the Book of Mormon is given in the testimonies of the Mormons who read this "greater part" and immediately left the Mormon Church.

Mormons preach that the *Book of Mormon* is the word of God, and that a person can get closer to God by reading the book. Mormons testify of the book's conversion power from their own experience. By reading just one book, their minds have been convinced of its truthfulness. But when asked to read the authentic *sealed part* that was published in 2004, the Mormons are unwilling and will not do what they ask everyone else in the world to do: read just one book with sincerity and with real intent.

Every Mormon, without exception, who has read *The Sealed Portion*, with a broken heart and contrite spirit, has left that religion. To date, this book is our most powerful scripture. It does what we had hoped it would do. It opens up the reader's mind to things that the reader never considered. It opens up the reader's mind so that we can introduce concrete ideas that will make this world a better place.

When we introduced our *Book of Mormon* through Joseph Smith, the Christian world rejected it. The early Mormons rejected the book as the source of teachings on which they *should* have established a new American religion. Once they did this, distraught dissenters found a way to kill Joseph Smith. The modern Mormons refuse to read our *Sealed Portion*. If they could find a way, they would also silence our modern True Messenger. We do everything within our power to protect him.

Chapter 9

Muslims and Mormons

We were the ones who initiated the beginnings of the Islamic religion of the seventh (7th) century in the Eastern Hemisphere, and also the beginnings of the Mormon religion of the nineteenth (19th) century in the Western Hemisphere. The evidences of our involvement in these two religions are the many glaring similarities between both.

As stated, in the spring of 1820, a young American teenager by the name of Joseph Smith, Jr. experienced the same *transfiguration* that Muhammad and that we had experienced. Similar to Muhammad (as we shared previously in his own words), Joseph was persecuted for his knowledge. How could a poor, uneducated youth know what the teenager Joseph Smith knew? How could a poor, uneducated Arab know what Muhammad knew?

Muhammad's First Vision of the angel Gabriel was Joseph's First Vision of God, the Father, and his Son, Jesus Christ. When Joseph first reported, under our direction, about what had happened to him as a teenager, he first reported an event that was similar to Muhammad's: that he had seen an angel from God.

Joseph told the story in several different ways until many of his original followers left him and accused him of being a fallen prophet. It was almost a decade after initially organizing a new American religion that we counseled Joseph to dictate "this history, to disabuse [correct] the public mind, and put all inquirers after truth in possession of the facts." Joseph Smith wrote, "I have been induced to write this history." We (the Real Illuminati™) "induced" or persuaded him.

We also "induced" Muhammad to present a story that fit in with both Jewish and Christian beliefs. The angel Gabriel appeared to Mary and announced the birth of Jesus. For the Jews, Gabriel was God's main messenger and revealer to God's prophets of God's will. So it made perfect sense to both Christians and Jews that the angel Gabriel would appear to God's prophet, Muhammad, and tell the people (through Muhammad) what God wanted them to do.

Joseph Smith was the American Muhammad. Mormonism was the first, new American Christian religion. Other than orthodox Protestantism, everything about the origins of Mormonism strangely resembles the origins of Islam, in almost every possible way. Mormonism is similar to Islam because the source of both religions' beginnings was the same: us, the Real Illuminati™.

Joseph reported to his followers that, after his First Vision, God's angel appeared to him from time to time and taught him. He reported,

> Accordingly, as I had been commanded, I went at the end of each year, and at each time I found the same messenger there, and received instruction and intelligence from him at each of our interviews, respecting what the Lord was going to do, and how and in what manner his kingdom was to be conducted in the last days.

Joseph Smith reported that these "interviews" in which he received "instruction and intelligence," started in September of 1823 and lasted for about four years.

That was when God was ready to reveal *His* "the word of God" to the world. Muhammad reported that, after *his* First Vision, he spent about four years receiving "instruction and intelligence" from the angel Gabriel. Then he was allowed to reveal "God's will" to the world. It took us four years of intense "instruction and intelligence" to prepare our two True Messengers for the role that they would play. We needed them to be able to introduce concepts into people's minds that might change the world.

In Muhammad's case, we didn't want to introduce our teachings through the written word, as we have explained. Everything that we wanted Muhammad to teach the people, he did orally. This was because both the Jews and the Christians believed it was the way that God revealed his will to his servants, the prophets. "Surely the Lord God will do nothing, but he revealeth his secret unto his servants the prophets" (Amos 3:7).

This passage in the Old Testament has been used continuously by both Islamic and Mormon missionaries. With it, they attempt to prove that *their* prophet is the only prophet to whom the true and living God will speak. We used this belief to our advantage. Muhammad's original words, those that came out of his mouth, were later misinterpreted and misrepresented by those who claimed to have heard them personally firsthand.

Our intended message, given through Muhammad, was about free will, compassion, kindness and tolerance. It was about establishing an economic policy that would afford people the opportunity to exercise their free will; in other words, establish "no poor among them." Islam is supposed to be about this. Mormonism is supposed to be about this.

"Islam" means to surrender to the will of God, supposedly through the teachings of God's only true prophet, Muhammad. If the Muslim people would have only surrendered themselves to being compassionate, kind, and tolerant, and established an economic policy that would have afforded people the opportunity to exercise their free will, then our mission through Muhammad would have been a success. But Muhammad's message did not succeed. Modern Islam is a humanity mess.

Islam is responsible for most of the unrest and instability caused by terrorism throughout our modern world. It is not Islam itself that is the problem—where Islam is the teachings of Muhammad. Rather it is the way that "God's" modern Islamic leaders interpret these teachings as they are accepted by the various Muslim sects throughout the world.

What does Allah (the name for "God" in Arabic) want His modern people to do? How are the teachings of Muhammad supposed to be interpreted and understood? The problem resides in different Islamic leaders claiming that they are the proper successors of the prophet Muhammad, and are the prophets to which Allah reveals His will to the people.

The Mormons (a title like Muslims) are members of the Church of Jesus Christ of Latter-day Saints (the name of the religion, like Islam). They believe that *their* only *true* God reveals His will through *their* prophet, seer, and revelator. The Mormons had the same problem with succession after their "prophet" died as the Muslims did before them.

Three months before his death, Muhammad made a speech at a place where he would often rest on his trips from Mecca to Medina. What he said in his speech has been interpreted in two very different ways. Here is what Islamic history reports that Muhammad said about his son-in-law, who was also his cousin, Ali ibn Abu Talib:

> O[h] people ... Of whomsoever I had been Master, 'Ali here is to be his Master. O[h] Allah, be a supporter of whoever supports him and an enemy of whoever opposes him and divert the truth to 'Ali.

If you're a modern Sunni, you believe that the prophet gave the leadership to <u>anyone</u> who belongs to the ancient Quraysh family. This is similar to the Jews believing that their Abrahamic family is more important than any other. If you're a modern Shia, you believe that Muhammad specifically gave the leadership to the whole community on one condition. It was only given through direct descendants of his great-grandfather, Hashim. Hashim's family line, as we explained above, were the poorer members of the Quraysh.

The majority of Muslims today are Sunni. Sunni traditions hold that the Caliph (the prophet, seer, and revelator) was chosen from among Muhammad's closest companions. This is similar to how Peter was chosen from among the twelve disciples, who were supposedly close companions of Jesus. The majority of Mormons today belong to the Church of Jesus Christ of Latter-day Saints. This Mormon majority believe that their Mormon Caliph is chosen from among the Twelve Apostles who lead their church.

This idea of the right to be a prophet was a topic of great discussion and debate among the early Mormons. They wanted a church that resembled what they believed the church was like when Jesus set it up. However, Jesus did *not* set up a church.

In the stories about Jesus, there were conflicts and arguments about who was the greatest apostle. According to the story, when Jesus' close companions argued among themselves about who was the greatest among them, Jesus set a child among them and told his disciples that a child was greater than all of them. "Except ye ... become as little children."

There is no mention of Jesus setting up an organized religion. In fact, there's a part of the story when a man, who was *not* part of Jesus' close companions, was preaching good ideas. Jesus' disciples tried to rebuke the man for teaching without permission. Jesus said, "He that is not against us is on our part."

Religious leaders want to hold on to their power and control over the people any way that they can. The Christian Mormons didn't pay any attention to Jesus' teachings, any more than the Muslims did to the teachings of Muhammad.

Some of Joseph Smith's early followers wanted Joseph to appoint twelve men as his apostles, as they thought Jesus' ancient Church had. But Joseph Smith refused to do this. He knew it would lead to the corruption, misinterpretation, and misrepresentation of the message he was asked to give the people, which he had received from us (the Real Illuminati™).

Instead of relying on Joseph conveying our message to the people only through his words, we wrote a new scripture that would relay our message to the people. We knew that the nineteenth (19th) century provided the means whereby what we wrote could be preserved. We knew that if we wrote the "word of God" the way that the people expected it to be given, then it couldn't be changed and misinterpreted the way that our message given through Muhammad had been. For this reason, we (the Real Illuminati™) wrote the *Book of Mormon* story and had Joseph Smith present it in a way that would convince the people that it was from God.

We intended for the story to include "the fullness of the everlasting Gospel." And because it was in written form in modern times, we figured that it could not be altered.

Then the early Mormons wanted their new church to include twelve apostles who would choose a Mormon Caliph (prophet, seer, and revelator) from among these twelve men. This man would have the authority and right to receive "God's will" for the people. We warned Joseph against this because of our experience with the Muslims.

Immediately after Muhammad's death, there was a debate (more of an intense argument) about who would succeed him. The young Islamic state was without a leader. It was reported that the Muslims were like sheep on a rainy night, huddled together and full of fear. Their beloved prophet had left no instructions on who should succeed him.

Within hours of his death, many of his close companions, who were popular figures in the Muslim community, got together to discuss what should be done. Some had been with Muhammad since the beginning and had accompanied him from Mecca. They prevailed. They chose Seddiq (the Mormons' Brigham Young) to succeed Muhammad.

None of Muhammad's immediate family were present during the debates. They were in mourning and were preparing Muhammad's body for burial. When Muhammad's son-in-law, Ali, who was married to

Muhammad's favorite daughter, Fatimah, heard that a new prophet had been appointed without his consent, he confronted those who had chosen Seddiq.

Realizing that he was outnumbered at the time, and still mourning his father-in-law's death, Ali reluctantly conceded to the choice. Many felt that Ali was a coward and had betrayed Muhammad's wishes that they believed their beloved prophet had proclaimed just three months before, which made Ali his rightful successor.

Joseph Smith and his brother Hyrum were murdered on June 27, 1844. Just a few days after that, Brigham Young announced that he had been chosen by the rest of the Mormon apostles to succeed Joseph Smith. Brigham Young was one of the men who had been chosen as one of the original Mormon Twelve Apostles. He was not chosen by Joseph, but by three of the most popular men in Mormon history: Oliver Cowdery, Martin Harris, and David Whitmer, also known as the Three Witnesses. They are the ones who actually chose the first, the original, Twelve Apostles.

Joseph Smith's wife, Emma Smith, and his mother, Lucy Mack Smith, were livid. These important women in Mormon history were still mourning their husband's and son's death, when Brigham came to their home. Brigham Young brought a companion, Heber C. Kimball, with him to announce that he (Brigham) was taking Joseph's place. You can imagine how this made these mourning women feel at the time. They kicked Young and Kimball out of their house and never again had anything to do with them.

About three months before he was murdered, Joseph Smith, as reported by many of his close companions, gave a blessing to his son, Joseph Smith, III. In this blessing, he conferred upon his son the mantle of prophet, seer, and revelator, if he were to die. No authentic document records this Mormon blessing, any more than any authentic document recorded what Muhammad supposedly said about Ali taking his reigns ("authority"). These accounts were oral testimonies given by many early Mormons and Muslims. Whether Muhammad had appointed Ali, or whether Joseph Smith had appointed his son to be their successors, our original message was completely distorted and ignored *because* of the controversy.

Both the Sunnis and Shias spend most of their time arguing (and even killing each other) over who was supposed to be God's succeeding messenger. Modern Mormons do not spend much time arguing over who was appointed to be God's prophet after Joseph Smith. There are many different modern-day Mormon sects. Each sect accepts their own different prophet. The Sunni-like Mormons (or rather, the larger majority) listen to no one but their Mormon Caliph, the President of their church. Whatever he says, *is* the will of the Mormon "Allah."

Here is the great irony of both Muslim and Mormon belief: The Holy Quran supposedly contains a fullness of the teachings of God. Likewise, Joseph Smith reported about how God's messenger (the Mormon Gabriel, whom the Mormons call "Moroni") delivered the *Book of Mormon* to him:

> He called me by name, and said unto me that he was a messenger sent from the presence of God to me, and that his name was Moroni; that God had a work for me to do; and that my name should be had for good and evil among all nations, kindreds, and tongues, or that it should

be both good and evil spoken of among all people.

He said there was a book deposited, written upon gold plates, giving an account of the former inhabitants of this continent, and the source from whence they sprang. He also said that the fulness of the everlasting Gospel was contained in it, as delivered by the Savior to the ancient inhabitants[.]

Nothing is more important to the Muslims than the Holy Quran. One would suppose that nothing would be more important to the Mormons than the book that was presented as being "written upon gold plates" and containing "the fullness of the everlasting Gospel ... as delivered by the Savior."

Both the Holy Quran and the *Book of Mormon* contain a special message. When read in context and with a sincere attempt to understand the message, it presents the idea of a perfect society. We gave the messages to Muhammad. We actually wrote a book for Joseph.

Our message is ignored. It is recited every day by the Muslims, but they do not live by its teachings. The Mormons have set the *Book of Mormon* aside, replacing it with whatever the Mormon Caliph ("prophet, seer, and revelator") proclaims is the "will of God."

We made our best attempts during the seventh (7th) and nineteenth (19th) centuries of this current dispensation of time, to influence the hearts and minds of humanity. We wanted them to incorporate into their societies and cultures the basic human principles of kindness, compassion, and most importantly, tolerance of others. These principles are set forth in different ways, according to the cultures in which we introduced them as the will of God (or Allah). These principles have not changed; but they are ignored.

Islam, which should be the most peaceful religion on Earth, has become the most violent. Mormonism, which should help people see the benefit of having a system of economic equality and equity (no poor among them), has become one of the richest churches in the entire world.

We have failed miserably through religion. Our efforts have only created more problems than good.

There is only one thing we have not yet tried: to tell the people of Earth the Real Truth.

This book introduces our desire to convince the people of Earth what they need to do to live in peace. If not, our *True Selves* will deliver them to the hardness of their hearts and the blindness of their minds to their own destruction ... because of religion.

Chapter 10

The Mortal Experience

We have presented the Real Truth that each person on Earth is actually connected to what could be perceived as that person's higher *Self or Higher Power*. We have explained that the deep spiritual feelings that cause a person to believe that there is a god outside of the mortal individual are real feelings. We have explained that this sensation is tied to your actual physical connection to your *True Self*—a connection that was made when you took your first mortal breath, and which will be disconnected when you take your last.

We have explained how religious leaders take advantage of these real feelings for their own benefit by setting up religions that define God and take away any control your *True Self* might hold over your mortal acts and thoughts.

We have explained how members of our group (the Real Illuminati™) have experienced a physical *transfiguration* that allows us more access to the reality that our *True Self* is experiencing. We have explained that upon gaining this access, our only desires have been, are, and will forever be to help encourage the rest of humanity to unite and change things upon Earth so that each person can begin to exercise that person's individual unconditional free will.

We have explained that this unconditional free will is the only way that our mortal Self can properly do what our *True Self* needs us to do in order to fulfill the purpose for which our *True Self* connected to a mortal body on Earth in the first place.

But what we haven't explained is *why* our *True Self* needs the mortal experience. In this book, we're going to touch upon it, but we're not going to give all the details of this. Eventually, if the world reacts in a positive way to this book, we will reveal the entire details of the importance of the mortal experience. For now, we need to explain a little bit about why this mortal experience is so important. This importance is why we do what we do.

Fortnite was created to serve a need for mortals (to relieve boredom). When the creators felt it was no longer serving the purpose for which it was created, at least to the degree that they expected, they destroyed it. Just as Fortnite serves a need of mortals, Earth life (the mortal experience) was created to fulfill a need of our advanced *True Selves*. Without life on Earth, life experienced as a higher, more advanced human will not be good.

But why does our *True Self* need the mortal experience? Why is it so important? Once mortals understand this, they can begin to create the best mortal experience for everyone.

To explain it, we need a little bit of science. Science is simply the way that some people view things on Earth and try to figure them out on their own. Throughout history, when people have had the chance to think about things, they've thought about *why* things are the way they are. Why are humans so different from all other living organisms? Why do humans think the way that they do about life and their experiences?

Science did not come first; philosophy did. Science is the experimentation on what we think about. Philosophy is just sitting around thinking about things.

But who has the time to just sit around all day thinking things or experimenting on their thoughts?

For most people on Earth, the largest part of their waking hours is spent just struggling to stay alive. They don't have the time or energy left to think about deep subjects that they don't know the answers to. So they pay others (in money, praise, or respect) to figure it out. Those to whom they pay are the false messengers who deceive and control humanity.

These are false messengers because nothing that they have done and none of the answers that they have provided, have done or do anything to make the world a better place for all people to live. These false messengers are the ministers of philosophy, religion, and science. We list these three main types of thought based on the way each has influenced humankind. Religion is a result of philosophy.

We have explained how religious thought and influence started. We have explained how the ancient philosophers (who had more time on their hands to sit around and think) created religion to justify their existence and create value for themselves. The other members of society were actually working to provide the needs of the community while these few sat around figuring out how to increase their personal value.

As societies developed and cooperation led to an easier life, more and more people were able to sit around all day and think about things. When the laboring class (who couldn't sit around all day thinking about things) was sufficiently satisfied (or rather, deceived) by a particular religion, other false messengers stepped in. Others who could afford to sit around all day thinking about things had to use their time to create value for themselves in ways other than religion. These began to experiment on their ideas and became scientists.

Consider modern scientific exploration and research. Who does it for free? Who does it to help all of humanity—for free? When a university receives a grant or gift of money to do scientific research, they need to *pay* those who sit around all day thinking and experimenting on past scientist's theories. So let's say that the scientists discover a cure for cancer. Is the university, or the sponsor that granted the money, going to give the cure for cancer to all of humanity for free?

In today's world, there are a lot of people who sit around all day in school thinking about theories that other people have already come up with and written down in a book. These hopeful scientists, of whatever degree, need to eat. For survival, they depend on those who produce the food, who *don't* have the time to sit around all day thinking.

The scientist is paid money provided by someone else (a sponsor) to buy this food. Obviously, the sponsor (the one providing the funds to the scientist) has more money than is needed to buy his or her own food. They give this extra money to the scientists. But where did this excess of money come from?

Sponsors (benefactors) make a lot of money off the laboring class (the majority). These people, most of the world today, can't sit around all day thinking about ideas. Instead, they are working twelve hours a day in the hot sun or in cramped factories, making sure the sponsor can make a profit. A portion of this profit can be given to the scientists, but not without the sponsor or benefactor *benefiting* from the investment (gaining value or making money from it).

This is why a cure for cancer will never be provided to all of humanity for free. This example of how universities' research departments operate IS the Real Truth. People go to college to get a degree because they don't want to have to work for twelve hours a day in the hot sun. They want to sit around all day and think about things, then experiment on what they thought about, in order to make some money. Their hope is that they will not be forced to do physical labor, but can use their brain to make money.

Consider an architect for example. An architect thinks about how to construct a building. They sat around all day in school learning what other architects had already thought about in the past. This is what they read in their architectural textbooks.

They studied about (for one of many examples) the architecture of ancient Greece and other cultures that developed great cities with big, extravagant buildings. After they have thought enough to satisfy those who are making them pay for their education (the colleges and universities), they get a piece of paper that proves that they sat around thinking a lot.

They show this piece of paper to someone who needs a building built. The paper proves that they supposedly know how to build buildings. Their value in society comes from the value that is placed on big buildings or spacious homes for those who have a lot of money to build them. This money is earned from the labor of hundreds of other people, who don't have time to sit around all day thinking about things.

Competition begins between architects and they struggle for value. They do everything that they can to convince a person who has extra money to build a new building, that *they* are the one who will do the best job designing it.

Perhaps one of the architects will argue that the university they went to was better than one that another architect attended. (Harvard is considered a much better school than a community college is.) So the architect who can convince the person with the money to build the new building, that his thoughts are best, wins.

Remember those hundreds of people working in the fields all day, who don't have the time to spend sitting around all day thinking about things? They don't need to compete for jobs. They don't need to convince anyone that what they think about in their head is better than what another laborer thinks about. Their boss doesn't want them to think. He or she (the benefactor) just wants them to work. This work consists of them doing the same thing every day, twelve hours a day, often 365 days a year.

Let's return to the sponsor who wants to make a good profit, so that they can grant a lot of money to a university's research department to find a cure for cancer. This sponsor doesn't pay much attention to the fact that the hot sun, the lack of rest, and the poor conditions in which the laborers work (the ones whom they're profiting from) ... actually *cause* the cancer that they're funding the research to cure. (And remember, the sponsor is also hoping to profit by giving a lot of money to the scientists at the university.)

Once the laborers get cancer, they have to use the money that their employer paid them for working, to the doctors to be cured of cancer. The cancer was caused by the employee working. This is another way that the few (the bosses) make even more money. In a sense, the laborers were actually working for free, because the money that the employer (benefactor) paid them was given back to the wealthy people, making them even richer.

Ninety-nine percent (99%) of the people in this world are laborers. Only one percent (1%), if that, are benefactors (sponsors). But at least one half of all the people in the world are doing everything in their power to *become* one of the one percent (1%), one of "the few." These are the Middle Class—those who want to sit around thinking and then go to the gym and get the same physical exercise that the laborers already get by working in the field.

The benefactors are those who sit around all day just thinking. The laborers are those who are forced to work all day long at the same job, doing the same thing. Without the Middle Class, there is no way that the one percent (1%), the few benefactors, can convince the 99% laborers that there is hope, *if* they just work hard enough.

This is because of what we explained in a previous chapter:

> Teetering on the edge of wealth and believing that they could become just as the wealthy, the middle class stood between the rich and poor: coveting [the rich and] fearing a return to [being poor—thus] validating [supporting] the actions of the rich, while pacifying the cries of the poor.

Whenever a few people have created value for themselves in a community without being forced to work, their lifestyle becomes easier. Because they don't have to work, the other members of the community have proof that it IS possible to live such a life of luxury. But the thing that motivates the poor more than anything else, something unique only to humans, is HOPE.

"Hope" is the intrinsic (basic internal) measure of our humanity, or better, that which we feel *can* be possible in spite of the improbabilities that seem to be part of our present experience. In other words, hope is the feeling inside of each of us that makes us *believe* that our individual life can be better than what it is. When we see others living a better life than our own, this convinces us, with actual evidence, that a good life is actually possible.

But what creates this feeling inside of us? What creates this "intrinsic (inner) measure" of things? What makes us think that we are just as good as everyone else and if someone else has a good life, then we deserve one too?

The answer is simple: the Real Truth about our connection with our *True Self.* It is this connection that gives us hope. It is this connection that creates the inner hope with which we measure what is a "good life" and what is not.

As our *True Selves*, we all live the good life on our advanced planets. It's just our reality. Our mortal Self is connected to our *True Self* for the purpose of existence, for the sake of our *True Self.* Because of this, however our *True Self* is living, then deep down inside of us (intrinsically, inner) we are going to *feel* that we should also be living "the good life" as a mortal.

We do not fault the few. How can we condemn or fault the few who are doing what their *True Self* wants them to do: live the good life? Above, we mentioned that we need a little bit of science to help explain this.

In a previous chapter, we explained about the energy of an advanced brain and that of a mortal brain. Here's what we explained:

Think of the energy that the brain produces <u>like</u> you would think of air pressure. An advanced brain has a higher level of pressure (energy). A mortal brain has a lower level of pressure (energy). If your advanced brain connects to your mortal brain, it is impossible for the air (or the energy) from a lower pressure chamber (your mortal brain) to enter where the higher pressure exists (into your advanced brain).

In order for the lower air pressure to enter a chamber that contains a higher level of air pressure, the air pressure in the higher chamber (advanced brain) must be lowered. This only happens when the rules are bent for a few, but not to the normal mortal person.

The mortal brain (chamber) is only capable of a certain level of energy (like air pressure). The advanced brain (chamber) that is connected to the mortal chamber must be lowered in order for the air to pass into the higher-pressured chamber. Again, this only applies to those of us within the game who understand the Real Truth about human existence.

The laws of mortal human nature on Earth do not allow the creation of a chamber of a mortal brain that can handle the higher pressure. Earth life was set up so that the higher pressure from the advanced brain could leach into the mortal brain of a lower pressure, but not vice versa. (For example, the air pressure inside an inflated balloon naturally rushes *out* of the balloon into the surrounding environment, but not the other way around.)

So, how does this leaching process take place? When you physically exert yourself, your mortal brain creates more pressure (more energy) by your brain forcing you to act in an environment. (Or it could be caused by whatever you are doing on Earth that creates more pressure.) Instead, when you can sit around all day and are not physically exerting yourself, the pressure in your mortal brain decreases.

In order for your mortal Self's brain to receive more information from the higher pressure level in your *True* Self's brain, you must lower the pressure of your mortal brain. The less physical exertion your mortal brain creates—the less physical work that you do, the more the energy from your *True Self* can leach into your mortal brain. The more energy your *True Self* can leach into your mortal brain, the more your mortal self can fulfill the purpose for which your *True Self* intended. This supports the purpose and importance of the mortal experience.

Chapter 11

Influencing the Renaissance

When the ancient Greeks established an economy in which a lot of people no longer had to do hard labor, more Greeks were able to lower the pressures of life. This allowed them to think more clearly and receive leached information from their actual *True Self*. Their *True Self* needed them to live the good life, so that they could exercise unconditional free will.

Many ancient Greeks were *born* into a world that forced them to work all day long. This didn't allow them to sit around all day and lower the pressure in their brains enough to exercise their individual free will. Our *True Selves* are forced to take whatever mortal body is available. If the only bodies that are available are those who have to work all day long, then those who choose to live on Earth must be born into those bodies. Being forced to work all day long does not allow the mortal self to completely fulfill the needs of the *True Self*. But then we can come back again, if we want (just like Fortnite), *reborn* into a new body. We can try again to live according to our *True Self's* needs. (The term "renaissance" actually means: rebirth.)

The Greek renaissance began in Athens, Greece about the seventh (7th) century BCE. Before the Greeks, the Egyptians, Sumerians, and other societies had experienced their own type of renaissance, but they were not as successful as the Greeks. The Muslim renaissance began in Baghdad of ancient Mesopotamia in the seventh (7th) century CE. The European renaissance began in the area of Florence, Italy in the fourteenth (14ᵗʰ) century CE. (At least this is how the historians recorded it.) There were many other cultures that also experienced their own type of renaissance, such as the Oriental renaissance, but these were influenced by the trade connections that they had with other developed cultures.

These time periods of *rebirth*, of *renaissance*, began when there were a few who could sit around all day thinking, painting, experimenting, composing music, and doing everything else that gave them a good life. But these were the few.

The majority had no choice but to create pressure in their brains that wouldn't allow energy from their *True Self* to leach through. The majority depended on the few to provide them with those things that produced good feelings in their brains. When another person is exercising unconditional free will, this gives those who are not, who cannot, the hope that they eventually *can* do the same.

Because of the good feelings that the majority do not *feel*, intrinsically (inside), they pay good money to those who make them feel good. This is why the majority love to listen to music. They love to see beautiful paintings. They love to be entertained. The majority are forced to work all day long. Thus, they are unable to create these good feelings themselves. So they pay whatever it takes to those who *can* provide them with these feelings ... with the hope.

This inner hope of the majority created the Middle Class each time a renaissance was successful. There was an ancient Greek Middle Class, which eventually led to an incredibly powerful Roman Middle Class. There was an ancient Muslim Middle Class. There was the Middle Class of medieval times (the European Middle Ages).

But none of these economic classes would have existed if not for the economic principles that created them. These economic principles are what sustain and support CAPITALISM.

Capitalism exists because of the *rebirth*, the *renaissance*, of a culture's ability to provide individual free will to a few of its citizens and, at the same time, create a hope for the rest. This same economic structure existed in Greece, in Rome, in Baghdad, and in the European cities of the Middle Ages. Capitalism provides the few with the ability to live a life of free will that is consistent with the needs of their *True Self*. This is true just for the few though.

Whenever Capitalism started in these renaissance cultures, we (the Real Illuminati™) would appear. We would travel wherever and whenever we found that a culture had started to provide people with the ability to exercise unconditional free will. Among these cultures, we sought out the few among the few who might help us introduce an economic system based on the *rebirth* of thinking (the *renaissance*), responsible for Capitalism.

It's not Capitalism that is bad. If used properly, Capitalism can give all people on Earth unconditional free will. But before this can happen, forced human labor has to be replaced with technology.

We know more about technology than all the scientific minds in this world combined. If the few would just listen to us to use the knowledge that we have for the right reason, we would gladly share what we know with the world—free of charge. For one example, if they would provide a cure for cancer free of charge to all people, we would share this knowledge with them.

But in our search for the few among the few, we have not been able to find any. In almost every case, those who we have found in a position of authority, who could influence the majority, would not help us. What politician today, what scientist today, what actor, what singer, what artist of any kind, will help us—for free? Their money and value come from keeping the majority in the dark and not allowing the majority to compete against them.

How many forced laborers are there who can sing or paint, and who, if given the chance, could compete with the few? And if the people ran their own government in the proper way, where the people retained their own power in how they were governed, what value would politicians have?

How many universities, pharmaceutical companies, and their respective research and development departments would provide their discoveries for free? Even if they were *given* the knowledge of how to cure disease or create energy in such a way that all of humanity could have as much electricity as each person needed, these would *not* provide them for free to others.

We mentioned how we approached Thomas Jefferson to help us introduce a new religion in the newly established United States. He refused to help us because of the political power and popularity that he was used to, that gave him his own good life.

We found a lot of thinkers in our search, whom we inspired with our knowledge. We do not sit around all day just thinking. We do not have to think; we already know. We spend all day, every day, searching for the few among the few who might help us give our knowledge to the world for free. We have yet to find one who has the clout and influence with the majority that is necessary in order to convince the majority to change the way life upon Earth is going.

Our searches in the past, among the many renaissance cultures, have led us to many people whom the world values as great thinkers, as great composers, great politicians, and great entertainers. Our interaction with them helped them to think about things differently and create philosophies that they otherwise would have never thought of, without our help.

How did we do this without revealing who we are and what we were trying to accomplish? We became their servants.

As one example, we became the servants of a man named Caneaus, who lived in Tunisia in the fourth (4th) century CE. Caneaus was the main writer and editor commissioned by the secret combination of politics and religion that produced the stories of Jesus. Caneaus needed to know what his target audience wanted to hear. His target audience was the poor. As his servants, or friends of his servants, we (the Real Illuminati™) were able to introduce the Sermon on the Mount.

In our new American scripture, we present a story about how we become the trusted servants of a powerful person in order to influence change. We invented a character to represent us and what we do and how we do it. We named him "Ammon."

Ammon wanted to help a culture of people living in a land that we called in our book, the "land of Ishmael." The people were under the power and control of a king named "Lamoni." Ammon was white. The people living in the land of Ishmael were dark-skinned people. The dark-skinned people hated the white-skinned people. Ammon wanted to help the dark-skinned people (who we named the "Lamanites") to not hate his culture of white-skinned people (that we named the "Nephites").

Ammon entered the land of the Lamanites and allowed himself to be arrested. Usually, as we wrote in our story, when the darker-skinned Lamanites encountered a lighter-skinned Nephite wandering in their land, they would arrest him, bind him, and carry him before their king:

> [I]t was left to the pleasure of the king to slay them, or to retain them in captivity, or to cast them into prison, or to cast them out of his land, according to his will and pleasure.

King Lamoni's servants brought Ammon to him. Ammon convinced the king that he wanted to "dwell among this people for a time; yea, and perhaps until the day I die." King Lamoni was pleased, but not convinced of Ammon's true intentions. To test Ammon, King Lamoni offered him one of his prettiest daughters. Ammon refused and said, "Nay, but I will be thy servant."

We presented this story in the way that we did to explain that the motives of a True Messenger or one of us (the Real Illuminati™), have nothing to do with sex, money, or anything else that most men desire out of life. The way we became the people's servants impressed whomever it was that we were trying to influence, who were usually men in authority and power.

In our story, Ammon's service to the king was responsible for changing an entire culture of people. Once Ammon had gained the trust of the people, he was able to introduce concepts that helped the dark-skinned Lamanite people see the white-skinned Nephites as their equals. We eventually renamed the "people of king Lamoni," the "people of Ammon." They were dark-skinned. They would not call themselves "Lamanites" however; nor would they call white-skinned people "Nephites": "And it came to pass that they called their names Anti-Nephi-Lehies."

These people in our story buried their weapons and made a promise to God that they would never use a weapon to harm another person. We presented these people as the most righteous people the world had ever known. They became an industrious people and there were no poor among them.

As we presented in our new American scripture story, we have always held on to the hope that we could help influence humanity to change its ways. It was our desire to help humanity become a united people that would never hurt each other. It was and is our hope that the people of Earth, of whatever race or color, would call themselves "Anti-American-European-Asians."

We have had a hand in the rebirth of various cultures throughout history, but have never succeeded in helping to influence the creation of the right one. The creation of the United States of America was our last hope.

Imagine that all the people on Earth finally develop and live in a perfect human world. Although improbable with the current forms of governments and socioeconomic systems (the way people interact with each other) that we allow to control our lives, we *can* accomplish this goal as a unified human race. It is not impossible to imagine a perfect world. In fact, it is what we hope for. "Hope," again, is the intrinsic measure of our humanity, or better, that which we feel can be possible in spite of the improbabilities that seem to be part of our present experience.

We wrote in the previous chapter titled "The Evil of Religion":

From the relatively little-known scientific "scripture" written by Robert Hooke and Max Planck to the more popular "prophecies" of Einstein and Oppenheimer, the first atomic bomb was created. This has destroyed many lives and has threatened the majority of the human race.

Not to be outdone by nuclear threat is the damage religion has created by the writings and actions *attributed* to the prophet Muhammad of the Islam faith and to Pope Urban II of Christianity. The Muslim conquests and Christian Crusades started "holy wars" (jihads) that have created both internal and external hatred between humans that linger even to this current day.

Human nature is responsible for these philosophies and the tragedies that they create. From the first time that mortals began to wonder about their existence and call upon an outside, invisible, unknown force to answer the questions about their reality, their innate human nature has deceived and misled them. The lingering questions of who we are, why we exist, where we came from, and where we go after we die, circumvent all other emotional activities of the human mind.

Humans have always struggled for the necessities of life, competing with each other for the limited resources of the earth. But this does not compare with the frustration they feel in their mind to find answers. The struggle *within* effects our emotional happiness the most. If one doesn't know the answers and cannot find them from within, it suggests that a person is flawed intellectually or implies that they are uneducated or ignorant in some other way.

Nevertheless, the questions still linger and the search for answers remain foremost in the mind. Is there someone else who might know the answers? Does there exist a messenger sent from the source of all knowledge to provide the answers? "If I don't know, then who does?"

Humans fight the thought that they might be stupid. They become discouraged when they can't find the answers themselves. A battle begins to take place in the mind. This enmity, this hostility, happens because each individual is connected to their True Self. There is a constant battle between the conscious mind (mortal Self) and the subconscious mind (True Self). Each person feels that they are just as good as everyone else. If they feel they don't understand something that someone else claims to understand, the human nature of equality compels their search for a better source of knowledge and understanding, or causes them to accept another's claims. This is how religions are formed.

Humans know that they are different from any other animal upon Earth. Subconsciously they recognize this difference, but they don't understand *why* or *how* they became so different. This inner battle, this opposition in one's mind (a "cognitive dissonance") compels them to figure it out.

In the search for the answers, the sincere and honest people (those who admit their lack of knowledge) become susceptible to the philosophies of men mingled with scripture. (This means that they become vulnerable or open to other people's opinions and ideas, as taught in religion, science, and philosophy.) This is because humans place an important value on knowing and having the innate questions about their existence answered. Because of this, when one can't figure out the answers for oneself, unscrupulous men (mostly men, some women) profess to have the answers for which the one is seeking.

Knowing of the intrinsic value and continual longing for answers, these false messengers set the price for the information they pretend to have. The one who provides the answer is allowed to bargain with the value of knowledge. And like everything else in a world in which "you can buy anything for money," the searcher will pay almost any price for the answers. They only need to be convinced that the answers are of worth.

The answers to life's questions have become a commodity that one can purchase, sell, trade, and negotiate for in other terms. The unscrupulous ones (those who are dishonest and corrupt) have become the merchants (the sellers) of knowledge. Because of the natural enmity (the battle for self-worth) within the human mind, these merchants do anything they can to protect the pretense (faking it) of their possession of knowledge. One way that they protect themselves and the group they have deceived into accepting their claim of knowledge—a group that gives them a lot of value—is by taking the treasures of the earth (money) they receive from the deceived group and financing armies to reign with blood and horror thereon; which is another way of explaining how people protect their family, community, and nation.

None of the gods created by the philosophies of men mingled with scripture are real. But their power over the human mind certainly is! From the moment that a sincere inquirer of truth, in submissiveness, kneels upon the earth and asks for guidance, "Oh God, hear the words of my mouth; Oh God, hear the words of my mouth. Oh God, hear the words of my mouth," that one becomes vulnerable to being answered by the only "god of this world": one's own mind.

But when the answers do not come from within oneself, the sincere seeker looks elsewhere. No matter how sincere and humble the person is (recognizing their inability to find the answers), they are forced to look for someone outside of their own consciousness to bring them the answers: "I am looking for messengers."

Throughout history, we have always been the messengers of Real Truth that the majority has been searching for. We have inspired the renaissance in any way we could. We don't blame the few for our failure to change this world into what it was meant to be. There is only one thing upon which we can properly place the blame. We blame the "the beast that was, and is not, and yet is," upon whose back humanity is riding.

Chapter 12

The Apocalypse

We have given the world proof that *we are who we say we are* (the Real Illuminati™). We have provided proof of what we have done. We have provided indisputable evidence that we know what we are doing. This sound evidence proves that we have done, are doing, and will do everything within our power to oppose superstition (illogical belief), religious influence over society, the abuses of government power, and obscurantism (the practice of deliberately preventing the facts or full details of a situation or event from becoming known by the majority). We will do everything within our power to put an end to the machinations (secret plans) of the purveyors of injustice (those who spread lies to create unfairness), to control them without impeding free will.

In a previous chapter, we wrote that the protection and support of individual free will is the overall goal of our group. We explained that throughout history, the secret combination of political, philosophical (which includes religion and science), and business powers has greatly diminished the ability of a person to pursue happiness according to that person's own desires; that wherever and whenever this secret combination has existed, so have we, in some form or another.

We have given the world proof that our efforts have been consistent and that we have not changed our goals, and that we are the same group today that we were in ancient times.

We used a lot of effort in the fifth (5ᵗʰ) century CE to make sure that the world would have this indisputable evidence of our existence and of our work. We were successful at getting the secret combination of political and religious powers at that time to include our *disclosure*, which gave the details about our work, albeit in a symbolic way.

We wrote this *disclosure* in a special way that has often been attributed to the idea of the existence of a secret group called the Illuminati. It has been said that we use secret signs, symbols, and codes. Those who say these things do not know us and are not a part of the Real Illuminati™. Those who have used different types of secret words, symbols, handshakes, clothing, etc. are the false Illuminati.

There were only two times in the current history of the world that we have used symbolic expressions to represent the Real Truth. The first time was in the fifth (5ᵗʰ) century CE.

We wrote our *disclosure* and were successful at getting it included in the Catholic Church's final canonization of religious authority: the New Testament of the Bible. Our *disclosure* became the final book: Revelation.

Written in Greek, the book's original name was "*Apocalypse*" (the Greek word for *disclosure*). The *apocalypse* has nothing to do with the end of the world as the false messengers and false Illuminati claim. It is a *disclosure* (description) of information about what has caused humanity's problems and the solutions that will preserve the human race.

In order to get the Catholic Church to include our *disclosure* in its canonized (sacred, holy) scripture, we were forced to use Jewish and Christian ideas and beliefs (which were the two main religions at the time) as the basis of our symbolism. We used symbolic expressions, of which <u>only we knew</u> the correct explanation. ONLY US!

Using religious symbolism helped us get our *disclosure* firmly imbedded in religious practice and belief. Our *disclosure* became part of the "word of God."

Basing the symbolism on things that <u>only we understood</u> protected our *disclosure* from the changes that religious leaders would later make to the New Testament, whenever it benefited the Church or its leaders. Because no leader or scholar in the Church understood what our secret symbolism meant, they couldn't change it. But they didn't *need* to change it. If they couldn't understand it, surely none of their followers could; so they left our *disclosure* alone.

For 1400 years, no person on Earth could properly unfold the symbolism of our *disclosure*, the book of Revelation. Many tried. There are countless explanations written by false messengers, erroneously trying to explain what Revelation's symbolism means.

A perfect example of this is that a large part of the world believes that the "Apocalypse" means the end of the world. The word "*apocalypse*" does not appear anywhere in the text of the book of Revelation. The word was not used anywhere except in the book's original title (which meant "disclosure").

Another general misconception is that the number "six hundred, three score and six" (666) refers to the devil—the devil being the "number of a man" mentioned. It has nothing to do with the devil. The deception has gained value for a lot of false messengers. We wrote in symbolism,

Here is wisdom. Let him that hath understanding count the number of the beast; for it is the number of a man; and his number is Six hundred threescore and six.

We knew that false messengers did not "have understanding." We knew that none of them would be able to solve the question about the "beast's number" being the "number of a man." Only <u>we</u> could explain what we meant. It took a True Messenger to finally unfold all of the symbolism found in our *disclosure*, our *Apocalypse*, in the book of *Revelation*.

The number "Six hundred threescore and six" is given in context with the few lines before it,

And that no man might buy or sell, save he that had the mark, or the name of the beast, or the number of his name.

"He that hath understanding" would know that whatever the "number" is, one must have it in order to "buy or sell."

We took all of our symbolism from the Christian Old Testament. To symbolize the "number of the man" that one must have in order to "buy or sell" in relation to "the beast," we borrowed from the passage that is in the modern King James Bible as 1 Kings 10:14: "Now the weight of gold that came to Solomon in one year was six hundred threescore and six talents of gold."

The "man" is Solomon. "His number" is gold. You cannot buy or sell without gold, which today is the equivalent of money.

The "beast" is Capitalism, nothing less and nothing more.

Capitalism is an economic system that can be explained as a human system of interaction. It's the way we interact with each other with things on Earth. Capitalism can be described as a socio-politico-economic system "that was and is not, and yet is." In other words, Capitalism is an *abstract* social, political, and economic system. This means that it exists in our thoughts as an idea, but does not have any physical or concrete existence.

It can be said that Capitalism is *marked* (evident) by the thoughts that originate in our forehead. The act that these thoughts create leaves a *mark* on society. Because we use our hands to perform the act, it can be stated that Capitalism's *mark*, the "mark of the beast," is both in our forehead and in our hand.

We do not need to provide any more explanation of the symbolism we used in creating our *disclosure*. We have already published a book associated with our work. The book is properly called, *666, the Mark of America—Seat of the Beast* (published 2006). There is no other book in this world—no other book before it, nor will there ever be any other book after it—that properly unfolds the symbolism of our ancient *disclosure* of those things that caused humanity's problems, and of the solutions that will save humanity.

While religion keeps us divided and unable to unite and solve our problems, nothing has caused more social, political, and economic problems than Capitalism. But if this "beast" is bridled and ridden properly, Capitalism can solve all of humanity's problems. We can explain how this can be done, if the world will listen to us. The proof that we are worth listening to is the profound evidence that no one on Earth was able to unfold the symbolism of our *disclosure* until we did it ourselves.

In this book, *The True History of Religion*, we mentioned that we had failed humanity in our efforts to help the people of Earth find the correct path to peace, harmony, and equality. But this is not the Real Truth. YOU have failed yourselves.

It is YOU whose pride and ego has put yourself above other people and does not allow the proper changes. You put your personal knowledge and beliefs, your opinions, above everyone else's. YOU are right and they are wrong. But to them, they are right and you are wrong—thus both of you feel you are right, which makes both of you wrong. On the other hand, the Real Truth is indisputable to everyone, at least to a rational mind that is supported by common sense.

You put your family above all other families on Earth. You put your nation above all other nations. And most destructive of all, you put your religion above all other religions.

We have failed in our efforts to help you think differently about religion. This is not because our efforts were not valiant and proper. This is because your pride and ego wouldn't allow you to accept that your opinions and beliefs are wrong; that your family is not any more important than any other family on Earth; that your nation is not any more important than any other nation on Earth.

In our new American scripture, we condemned the inequality between you and others, especially the economic inequality. We gave you permission to pursue wealth through Capitalism, but with one purpose and one purpose only. This was "for the intent to do good—to clothe the naked, and to feed the hungry, and to liberate the captive, and administer relief to the sick and the afflicted."

Instead, you became the richest religion, per capita, in the entire world. Your secret combinations of political, religious, and business powers do nothing but put yourself, your family, and your nation above all other people, above all other families, above all other nations.

The new scripture we provided for you even told you how important the book of Revelation should be to your religion. We presented a story where one of the main characters was being shown, "many things which have been ... also ... *concerning* the end of the world." He was not allowed to write everything that he saw. He was told by an angel of God (according to the story) that the book of Revelation would reveal everything, up until the end of the world. (As previously mentioned, when read correctly and in context, the book of Revelation does not foretell of the end of the world. It speaks of "renewing" the Earth.)

We presented the significance of "the Father's work" commencing when the book of Revelation was finally unfolded to the entire world. But you have completely ignored Revelation. You have completely ignored the intent behind the new American scripture that we provided for you.

You ignore these things because nothing is more important to you than your erroneous beliefs, your family, your nation, and your corrupted religion.

We speak to YOU to whom we delivered the new scripture, because it was in the United States that we believed that there was a hope to inspire humanity to use its power and influence in the world "for the intent to do good." But you have done nothing that is good.

You build churches and great temples that rise high above the earth, and into which you enter with this pride that you have in your religion, in your family, in your nation, and in your god.

We wrote of YOU, not about an ancient people, but about YOU in the latter days, in YOUR OWN SCRIPTURE:

> Behold, I speak unto you as if ye were present, and yet ye are not. But behold, Jesus Christ hath shown you unto me, and I know your doing.

> And I know that ye do walk in the pride of your hearts; and there are none save a few only who do not lift themselves up in the pride of their hearts, unto the wearing of very fine apparel, unto envying, and strifes, and malice, and persecutions, and all manner of iniquities; and your churches, yea, even every one, have become polluted because of the pride of your hearts.

> For behold, ye do love money, and your substance, and your fine apparel, and the adorning of your churches, more than ye love the poor and the needy, the sick and the afflicted.

> O ye pollutions, ye hypocrites, ye teachers, who sell yourselves for that which will canker, why have ye polluted the holy church of God? Why are ye ashamed to take upon you the name of Christ? Why do ye not think that greater is the value of an endless happiness than that misery which never dies—because of the praise of the world?

Why do ye adorn yourselves with that which hath no life, and yet suffer the hungry, and the needy, and the naked, and the sick and the afflicted to pass by you, and notice them not?

Yea, why do ye build up your secret abominations [your secret combinations] to get gain, and cause that widows should mourn before the Lord, and also orphans to mourn before the Lord, and also the blood of their fathers and their husbands to cry unto the Lord from the ground, for vengeance upon your heads?

Behold, the sword of vengeance hangeth over you; and the time soon cometh that he avengeth the blood of the saints upon you, for he will not suffer their cries any longer.

We influenced the scientific thought and mathematical calculations that allowed scientists to create nuclear weapons. The intent was not to destroy humanity, but to deter false messengers. We wanted to help the United States of America set up a government "for the intent to do good." This was our intent.

In introducing a new scripture, it was our "intent to do good." The new scripture was not about European Americans. They were greatly prospering in the "land of promise" that they stole from the native American peoples (including all of North and South America). The new scripture was about setting up a "New Jerusalem," a new place of hope, a place of economic equality for the native American people.

But you hijacked this new scripture, completely ignored its message, and used it to put yourselves, your religion, your family, and your nation above all others upon Earth. You even built a multi-billion (not million, but billion) dollar mall near your religious headquarters. You are the epitome of those with the mark of the beast on their hands and in their foreheads.

We did not state in vain that, "the sword of vengeance hangeth over you; and the time soon cometh that he avengeth the blood of the saints upon you, for he will no longer suffer their cries."

"Avenge" means to inflict harm in return for an injury or wrong done to another. What have you done to the native American peoples of the Western Hemisphere? Do they not suffer and cry and plead at your nation's border to enter the land of their ancestors, which you stole from them? And what about your economic policies? What have they done to these people?

Do you know what the "wrath of God" is? You do not! You know only what your false messengers tell you. Let us tell you what we meant by this.

It was connected to the "power of God," as we presented it in our story, which gave your nation the power of nuclear weapons. God gave you this power in order to build up a righteous nation, to influence a righteous world, but you have misused this power. It is a power that can be used by any who have the knowledge that it takes to put together the right combination in the right forms of the right elements.

People all around the world are crying and suffering because of the beast that you ride indiscriminately (aimlessly) throughout the world, in support of your pride and ego.

We have explained to you how to bridle this great beast (Capitalism) and use it for the benefit of all people on Earth. We have given you this information through a new idea for government, which we have presented as **The Humanity Party®**, the only political party in the world with the "intent to do good—to clothe the naked, to feed the hungry, and to liberate the captive, and administer relief to the sick and the afflicted." But you have ignored us.

The cries of the poor throughout the world can no longer be ignored by the *true god*, by our *True Selves*. Hope is fading. Love is sickened. Free will no longer exists. For the majority, nothing their mortal brain is doing, no experience their brains are producing on this earth, is useful to their *True Self*—to the only true god that sits on the throne within each person's brain.

Their cries will no longer be ignored.

We have done everything in our power to keep nuclear knowledge and technology out of the hands of the terrorist groups that have risen out of poverty and inequality throughout the world. We have explained that we, the Real Illuminati™ and our True Messengers, are servants of the *true god*. We only seek for peace, harmony, and equality. But

> the sword of vengeance hangeth over you; and the time soon cometh that he avengeth the blood of the saints upon you, for he will not suffer their cries any longer.

If we fail to inspire change and get the people of this world to come down out of that great and spacious building of their pride that puts one family, one nation, and one religion above another, our mandates from the *true god* are clear: **Do no more.**

It is just a matter of time before the same terrorist groups that have the knowledge to create an Improvised Explosive Device (IED), will possess the same nuclear knowledge and capability. Uncontrolled in their passion to be avenged of the poverty and inequality by which the great beast of Capitalism threatens them, they will fight the beast—soon with *nuclear* IEDs.

We know how to bridle the beast. We know how to control it. We know how to keep it from hurting others. We know how to create peace, harmony, and equality upon this earth. We know how to do it without impeding free will.

If your *True Self* (your god) wants you to be wealthy, we know how to help you ride the beast with the right saddle, bridle, and spurs. We will help you do it, but "for the intent to do good—to clothe the naked, and to feed the hungry, and to liberate the captive, and administer relief to the sick and the afflicted."

Our intent is to do good. But we can only do what we can do. Our work is a great work. It is a marvelous work and a wonder among the people of Earth. It will either convince them of peace and life, or deliver them to the hardness of their hearts and the blindness of their minds. In this way, they will be brought down into captivity and also into destruction, both temporally and spiritually—temporally through the destruction of terrorists, who are like "young lions."

We wrote of these terrorists in our new American scripture:

And my people who are a remnant of Jacob shall be among the Gentiles, yea, in the midst of them as a lion among the beasts of the forest, as a young lion among the flocks of sheep, who, if he go through both treadeth down and teareth in pieces, and none can deliver. Their hand shall be lifted up upon their adversaries, and all their enemies shall be cut off.

The adversaries of the few are the majority. The enemy of the people is the unbridled and out-of-control great beast that carries the few on its back, trampling the masses under the weight of its massive body. If this beast is not bridled and controlled, our Fortnite, our Earth, will be destroyed.

Help us help humanity. Allow us to help humanity. We will come forth, if wanted. We will show the world what <u>must</u> be done, what <u>can</u> be done.

First, the world must accept ...

The True History of Religion.

Lightning Source UK Ltd.
Milton Keynes UK
UKHW021851090223
416682UK00014B/1575